Simon Henry Gage

The Microscope

An Introduction to Microscopic Methods and to Histology

Simon Henry Gage

The Microscope
An Introduction to Microscopic Methods and to Histology

ISBN/EAN: 9783337397562

Printed in Europe, USA, Canada, Australia, Japan

Cover: Foto ©berggeist007 / pixelio.de

More available books at **www.hansebooks.com**

THE MICROSCOPE

AN
INTRODUCTION TO MICROSCOPIC METHODS AND TO HISTOLOGY

BY SIMON HENRY GAGE
PROFESSOR OF MICROSCOPY, HISTOLOGY AND EMBRYOLOGY IN CORNELL UNIVERSITY, AND THE NEW YORK STATE VETERINARY COLLEGE.

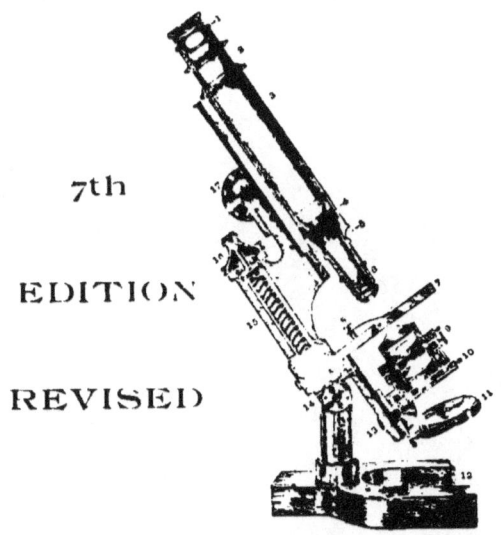

7th

EDITION

REVISED

COMSTOCK PUBLISHING COMPANY
ITHACA, NEW YORK
1899

PREFACE TO THE SEVENTH EDITION.

IN this edition pp. 17-18, 49-52, 103-104, 165-168, that is, the matter pertaining to *numerical aperture*, refraction, the *filar micrometer*, *imbedding* and *sectioning* by the *paraffin method*, have been rewritten. Some errors are noted, and additional or more advantageous methods suggested for some of the sections. The title page has been slightly altered in wording to make it more fully indicate the aim of the book.

ITHACA, Jan. 11, 1899.

Corrections.—Pp. 33, 39, 46, for Pleurasigma, read Pleurosigma. P. 97, second paragraph, change No. 4 to No. 3. P. 159, top line, change carbol-xylene to castor-xylene.

Additions.—P. 145, § 227: If the proportions of acid and dichromate are greater, this cleaning mixture will be more efficient. The following from Dr. G. C. Caldwell's laboratory guide is excellent. Potassium dichromate 40 grams; water 150 c.c.; sulphuric acid 230 c.c. For preparing this mixture, an iron kettle lined with heavy sheet lead has proved both satisfactory and economical.

Pp. 154, 177, §§ 245, 308. Formaldehyde dissociator of the strength of 2 c.c. of formalin, etc., to the liter, has been found more satisfactory than 5 c.c. to the liter.

Pp. 160, 166, §§ 261, 275. As pointed out in 1891 (Proceed. Amer. Micr. Soc., Vol. xiii, p. 82) it is of great advantage to albumenize the slides on which collodion sections are to be mounted. This is done by placing the cleaned slides in a jar of egg albumen 1 to 200 of water. This should be filtered before use. After half an hour or more the slides are removed, stood on end on a towel or blotting paper and allowed to dry; they can then be stored in glass jars and are ready to use at any time. From such albumenized slides the collodion sections, fastened as directed in § 261, will very rarely become detached even with repeated manipulation. Such albumenized slides are excellent for use with paraffin sections. If the sections are to be extended with warm water (§ 274) this is a much preferable method to that with Mayer's albumen.

P. 160, §§ 257-259. When objects are imbedded in a box for collodion sections they must ordinarily be fixed to some holder before sectioning. After the collodion is hardened in chloroform and clarified, remove the paper box, absorb the castor-xylene on the surface, trim the end and put some fresh, thick collodion on the cork or other holder. Press the imbedded tissue firmly against the holder. Within two minutes it will be firmly cemented and one may proceed at once to clamp the holder in the microtome and commence cutting.

ADDITIONS.

P. 160, § 260. For handling collodion sections the better quality of white tissue paper sold by stationers has been found excellent. It is worth while to have it cut into pieces about 60 × 50 mm.

Pp. 171, 179, 292, 296, 311. For a laboratory it has been found advantageous and economical to furnish the students with gummed labels for their preparations. The form used is shown in the accompanying figure. Labels of this kind can be bought in five thousand lots for 35 to 40 cents per thousand.

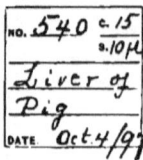

Also for temporary storage, and for sets of preparations to be issued to students, inexpensive slide drawers fitting the lockers have been prepared. The accompanying figures indicate the consturction. These in sizes which will hold 50 slides (30 × 43 cm.) cost from $12 to $15 per 100.

P. 175, § 258. *Grades of Alcohol.* It has been found by careful tests that quite accurate percentages of alcohol may be obtained by mixing water and alcohol as follows: Pour alcohol into a graduate until the volume of alcohol corresponds to the desired percentage. Add water until the volume in cubic centimeters corresponds to the original percentage of the alcohol used. For example, to get 67% from 95% alcohol, pour 67 c.c. of 95% alcohol into a graduate, and add sufficient water to bring the volume up to 95 c.c. For 50% alcohol from 75%, put 50 c.c. of 75% alcohol in a graduate, add sufficient water to make the volume 75 c.c. From the change in volume it does not answer to mix given volumes of water and alcohol in these cases. In the first case, if one mixed 75 c.c. of 95% alcohol and 20 c.c. of water the resulting mixture would be over 75%; but if sufficient water is added to bring the volume back to the original percentage more than 20 c.c. of water is added, that is enough more to compensate for the shrinkage, and the result is approximately accurate.

P. 175, § 299. In case preparations are to be kept some time in alum water, 2% of chloral hydrate should be added to prevent mold.

P. 216, § 360. For lettering diagrams, the so-called "easy sign markers" have proved very satisfactory.

PREFACE TO THE SIXTH EDITION.

THE rapid advance in microscopical knowledge, and the great strides in the sciences employing the microscope as an indispensable tool, have reacted upon the microscope itself, and never before were microscopes so excellent, convenient and cheap. Indeed, the financial reason for not possessing a microscope can no longer be urged by any high school or academy, or by any person whose profession demands it.

Naturally, to get the greatest good from instruments, tools, or machines of any kind, the one who uses them must understand the principles upon which their action depends, their possibilities and limitations.

That the student may acquire a just comprehension of some of the fundamental principles of the microscope, and gain a working acquaintance with it, this book has been prepared. It is a growth of the laboratory, and has been modified from time to time to keep pace with optical improvements and advancing knowledge.

This edition has been largely rewritten. Many new figures and about ninety pages of new matter have been added, and it is hoped that the student will find it a real help in his efforts to become master of the modern microscope.

SIMON HENRY GAGE,
CORNELL UNIVERSITY.

October 31, 1896.

PREFACE TO THE FIFTH EDITION.

THIS edition has been enlarged nearly one-half by the elaboration of the matter in the previous edition, and by the addition of a wholly new chapter on photo-micrography and on photographing natural history objects in a horizontal position with a vertical camera. The figures have been distributed in the text, and many new ones added.

It is hoped that the book as it now appears may, while remaining strictly elementary, still more fully meet the needs of those who wish to use the microscope for serious study and investigation. The aim has been to produce a book for beginners in microscopy, such as the author himself felt sorely the need of when he began the study. This purpose has been strengthened and furthered by noting the difficulties of the various classes that have used the work and aided in its evolution during the last fifteen years.

The author wishes to acknowledge the aid rendered by the various Optical Companies for information freely given, and for the loan of cuts and instruments (Bausch & Lomb Optical Co., Gundlach Optical Co., Queen & Co., and all the opticians mentioned in the table of tube-length, p. 10). I feel under special obligation to my various classes for the enthusiasm and earnestness with which they have followed the instructions in the book, to my colleagues, Professor Wilder and Instructors Hopkins and Fish for suggestions, to Mrs. Gage for criticising the manuscript, reading proof, preparing the index and the original figures, to Dr. A. C. Mercer for aid in preparing the chapter on photo-micrography, to Dr. M. D. Ewell for information and for the loan of apparatus, and finally, to many other friends who have used the previous editions, and have made suggestions whereby it is hoped the present edition is greatly improved.

I would like to repeat a part of the preface to the third and to the fourth editions, and to call especial attention to the address of the Hon. J. D. Cox at the recent meeting of the American Microscopical Society: "A plea for systematic instruction in the technique of the microscope at the university," in the Proceedings for 1893.

Extract from the preface of the fourth edition :

"The author would feel grateful to any person who uses this book if he would point out any errors of statement that may be discovered, and also suggest modifications which would tend to increase the intelligibility, especially to beginners."

From the third edition :

"It is thoroughly believed by the writer that simply reading a work on the microscope, and looking a few times into an instrument completely adjusted by another, is of very little value in giving real knowledge. In order that the knowledge shall be made alive, it must be made a part of the student's experience by actual experiments carried out by the student himself. Consequently, exercises illustrating the principles of the microscope and the methods of its employment have been made an integral part of the work.

"In considering the real greatness of the microscope, and the truly splendid service it has rendered, the fact has not been lost sight of that the microscope is, after all, only an aid to the eye of the observer, only a means of getting a larger image on the retina than would be possible without it; but the appreciation of this retinal image, whether it is made with or without the aid of a microscope, must always depend upon the character and training of the seeing and appreciating brain behind the eye. The microscope simply aids the eye in furnishing raw material, so to speak, for the brain to work upon.

"The necessity for doing a vast deal of drudgery, or 'dead work,' as it has been happily styled by Professor Leslie, before one has the training necessary for the appreciation and the production of original results, has been well stated by Beale:

"'The number of original observers emanating from our schools will vary as practical work is favored or discouraged. It is certain that they who are most fully conversant with elementary details and most clever at demonstration, will be most successful in the consideration of the higher and more abstruse problems, and will feel a real love for their work which no mere superficial inquirer will experience. It is only by being thoroughly grounded in first principles, and well practiced in mechanical operations, that any one can hope to achieve real success in the higher branches of scientific enquiry, or to detect the fallacy of certain so-called experiments.'"

SIMON HENRY GAGE,
CORNELL UNIVERSITY,
ITHACA, New York, U. S. A.

February 12, 1891.

CONTENTS.

CHAPTER I.
§ 1- 55—The Microscope and its Parts—Demonstration of the Function of each Part. Figures of Laboratory Microscopes, ... 1- 32

CHAPTER II.
§ 56-119—Lighting and Focusing, Manipulation of Dry, Adjustable, and Immersion Objectives; Care of the Microscope and of the Eyes, ... 33- 79

CHAPTER III.
§ 120-144—Interpretation of the Appearances under the Microscope, . . . 80- 91

CHAPTER IV.
§ 145-167—Magnification of the Microscope; Micrometry, ... 92-108

CHAPTER V.
§ 168-178—Drawing with the Microscope, ... 109-119

CHAPTER VI.
§ 179-218—Micro-spectroscope and Micro-polariscope; Use and Application, ... 120-139

CHAPTER VII.
§ 219-322—Slides and Cover-glasses; Mounting; Isolation; Sectioning by the Collodion and Paraffin Methods; Labeling and Storing Microscopical Preparations; Preparation of Reagents; Experiments in Micro-chemistry, ... 140-182

CHAPTER VIII.
§ 322-351—Photo-micrography and the Photography of Natural History Specimens in a Horizontal Position with a Vertical Camera, . 183-209

APPENDIX.
§ 352-370—The use of Abbe's Test-Plate and Apertometer, ... 210
Testing Homogeneous Liquids; Experimental Determination of the Equivalent Focus of Objectives and Oculars; Preparation of Diagrams; Preparation of Drawings for Photo-engraving, ... 213-219

BOOKS AND PERIODICALS 220-225
INDEX 227-237

LIST OF ILLUSTRATIONS.

The author extends grateful acknowledgments to the opticians and others who have loaned cuts for this edition. The source of each figure is given when borrowed. The other figures were drawn expressly for this work by Mrs. Gage. The frontispiece was drawn by Mr. Gutsell, of the University Art Department.

FIG.		PAGE.
	Frontispiece	
1-9.	The principal axis and center of various lenses	2
10-11.	Principal focus with converging and diverging lenses	3
12.	Chromatic aberration	4
13.	Spherical aberration	4
14-15.	Real and virtual image with convex lenses	5
16.	Simple microscope and eye of observer	6
17.	Tripod magnifier	7
18.	Achromatic triplet The Bausch & Lomb Opt. Co.)	7
19.	Lens-holder (The Bausch & Lomb Opt. Co.)	8
20.	Dissecting microscope (The Bausch & Lomb Opt. Co.)	9
21.	Principle of the compound microscope	10
22.	Dry objective	11
23.	Immersion objective	12
24.	Tube-length	15
25.	Tube-length when nose piece and ocular micrometer are used (Zeiss' catalog, No. 30)	16
26.	Angular aperture	17
27-29.	Dry and immersion objectives (Ellenberger)	18
30.	Section of Huygenian ocular for eye-point	22
31.	Compensation oculars (Zeiss' catalog, No. 30)	24
32.	Projection oculars (Zeiss' catalog, No. 30)	25
33-34.	Ocular micrometer with movable scale (Bausch & Lomb Opt. Co.) . .	25
35.	Ocular screw micrometer (Zeiss' catalog, No. 30)	26
36.	Triple nose-piece or revolver (Queen & Co.)	27
37.	Size of field with various objectives and oculars	29
38.	Principle of the simple microscope (Fig. 16 repeated)	31
39-40.	Dry and immersion objectives (Figs. 22-23 repeated)	34
41.	Achromatic condenser (Zeiss' catalog, No. 30)	41
42-43.	Image of diaphragm in centering	42
44-45.	Centering the source of illumination on the object	43
46-47.	Aperture of condenser (from Nelson)	43
48-51.	Abbe condenser, central, oblique and dark ground illumination	47
52.	Lamp and bull's eye condenser	49
53-55.	Refraction diagrams (from Carpenter-Dallinger)	50
56.	Aberration produced by the cover-glass (Ross)	52

LIST OF ILLUSTRATIONS.

57. Cover correction by changing tube-length 54
58. Screen for face and microscope 56
59. Ward's eye shade (Cut loaned by Queen & Co.) 59
60. Double eye-shade 59
61-63. Marker, sectional view (Proc. Amer. Micr. Soc., 1894) 64
64-66. Specimens showing the use of the marker 64
67. Krauss' method of marking objectives on a nose-piece (from Dr. Krauss, see Proc. Amer. Micr. Soc., 1895) 65
68. Removable mechanical stage (Leitz catalog) 65
69. Removable mechanical stage (Bausch & Lomb Opt. Co.) 65
70. Zeiss' Microscope Ia with mechanical stage (Zeiss' catalog, No. 30) ... 66
71. Watson & Sons, Edinburgh, student's microscope (Watson & Sons catalog) 67
72. Nachet et Fils microscope No. 4 with movable stage (cut loaned by the Franklin Educational Co.) 68
73. BB Microscope of the Bausch & Lomb Optical Co. (B & L) 69
74. Reichert's microscope IIIb (cut loaned by Richards & Co.) 70
75. Queen & Co.'s microscope II of the continental pattern (Q. & Co.) ... 71
76. Leitz' microscope Ib (cut loaned by Wm. Krafft) 72
77. Ross eclipse microscope (cut from Walmsley, Fuller & Co.) ... 73
78. AA. Microscope of the Bausch & Lomb Optical Co. (B. & L.) .. 74
79. Beck's star microscope (cut loaned by Williams, Brown & Earle) 75
80. Zentmayer's clinical microscope (Zentmayer) 76
81. Zentmayer's microscope, No. V (Zentmayer) 76
82. Leitz' demonstration microscope (from Wm. Krafft, N. Y.) 77
83. Leitz' microscope IV (from Wm. Krafft, N. Y.) 77
84. Queen & Co.'s acme microscope, No. IV (Q. & Co.) 78
85. McIntosh's scientific microscope, No. 2 (McIntosh Battery Co.) ... 79
86. Letters mounted in stairs to show order of coming into focus ... 82
87. Putting on a cover-glass 84
88. Oil and air bubbles 85
89. Glass rods in optical section 86
90. Double contour 87
91. Micrometer with ring to facilitate finding the lines 94
92. Wollaston's camera lucida 95
93. Geometrical diagram showing size of object and image 96
94. Image and object with differing tube-length 96
95. Standard distance for magnification with Wollaston's camera lucida ... 98
96. Standard distance for magnification with the Abbe camera lucida 98
97. Preparation of blood corpuscles with ring around a group 101
98-99. Ocular micrometer (Figs 33-34 repeated) 103
100. Ocular screw micrometer (Fig. 35 repeated) 104
101. Lines of stage and ocular micrometer in getting the valuation of the ocular micrometer 107
102. Abbe camera lucida with 45° mirror 110
103. Geometrical figure going with Fig. 102 110
104. Ocular showing eye-point (Fig. 30 repeated) 110
105. Wollaston's camera lucida (Fig. 92 repeated) 111
106. Abbe camera lucida with 35° mirror 114
107. Geometrical figure going with Fig. 106 114
108. Upper view of the prism of the Abbe camera lucida 114

109. Quadrant attached to the mirror of the Abbe camera lucida	114
110. Inclined microscope with the Abbe camera lucida	115
111. Drawing board for the Abbe camera (The Bausch & Lomb Opt. Co.)	116
112. Micrometer lines indicating the scale of a drawing	118
113. Longisection of the Abbe micro-spectroscope (cut loaned by the Bausch & Lomb Opt. Co.)	121
114. Slit mechanism of the micro-spectroscope (from B. & L.)	121
115. Various spectrums	122
116. Absorption spectrum of hemoglobin, etc. (Gamgee & McMunn)	124
117. Section of the micro-spectroscope	126
118. Prism showing apparent reversal of colors	126
119. Section of a micro-polariscope	126
120. Micrometer calipers (Brown & Sharp)	143
121. Cover-glass measurer (The Bausch & Lomb Opt. Co.)	144
122. Zeiss cover-glass measurer (from Zeiss' catalog)	145
123. Putting on a cover-glass (Fig. 87 repeated)	146
124. Needle holder (Queen & Co.)	146
125. Turn table (Queen & Co.)	148
126. Centering card	149
127. Anchoring a cover-glass	150
128. Irrigation, staining, etc., under the cover	150
129. Moist chamber for fibrin, blood corpuscles, etc. (from Proc. Amer. Micr. Soc., 1891)	151
130. Adjustable lens holder (Leitz, cut from Wm. Krafft)	155
131. Adjustable lens holder (The Bausch & Lomb Opt. Co.)	156
132. Preparation vials (Proc. Amer. Micr. Soc., 1895)	159
133. Pipette for stains, etc. (Whitall, Tatum & Co.)	161
134. Waste bowl with rack and funnel (cut loaned by Wm. Wood & Co.)	162
135. Round aquarium for waste bowl, rinsing jar, etc. (Whitall, Tatum & Co.)	162
136. Glass box for cleaning slides and covers (Whitall, Tatum & Co.)	162
137. Balsam bottle	164
138. Serial section slide, showing order of arranging sections	170
139. Writing diamond (Queen & Co.)	173
140. Drawer of cabinet for slides (Proc. Amer. Micr. Soc., 1883)	174
141. Cabinet for microscopical specimens (Proc. Amer. Micr. Soc., 1883)	174
142. Czapski's iris diaphragm ocular (Zeiss' catalog, No. 30)	181
143. Walmsley's large photo-micrographic camera (from Mr. Walmsley)	186
144. Leitz' vertical photo micrographic camera (from Wm. Krafft)	188
145. Projection oculars (Fig. 32 repeated)	189
146. Walmsleys autograph camera in a vertical position (from Mr. Walmsley)	190
147. Same in horizontal position	192
148. Vertical photo-micrographic camera (the Bausch & Lomb Opt. Co.)	193
149. Zeiss' 70 millimeter projection objective (from Zeiss' photo-micrographic catalog)	195
150. Focusing screen	195
151. Perigraphic photographic objective (The Gundlach Opt. Co.)	196
152. Zeiss anastigmatic photographic objective (from the Bausch and Lomb Opt. Co.)	196
153. Focusing glass (from the Gundlach Opt. Co.)	197
154. The tripod as a focusing glass	198

155. Engraving glass (The Bausch & Lomb Opt. Co.) 198
156. Bausch & Lomb's chain lens-holder for use with a dissecting lens, holding an engraving glass, etc. (B. & L.) 199
157. Bull's eye and lamp (Fig. 52 repeated) 200
158. Zeiss' vertical photo-micrographic camera (Zeiss' catalog) 202
159. Rack for drying negatives (Rochester Opt. Co.) 203
160 161. Sections of the head and brain of Diemyctylus (Mrs. Gage, from the Wilder Quarter Century Book) 203
162. Vertical photographic camera for picturing brains and other preparations in a horizontal position 207
164. Abbe's test plate 211
165. Abbe's apertometer (Zeiss' catalog) 212

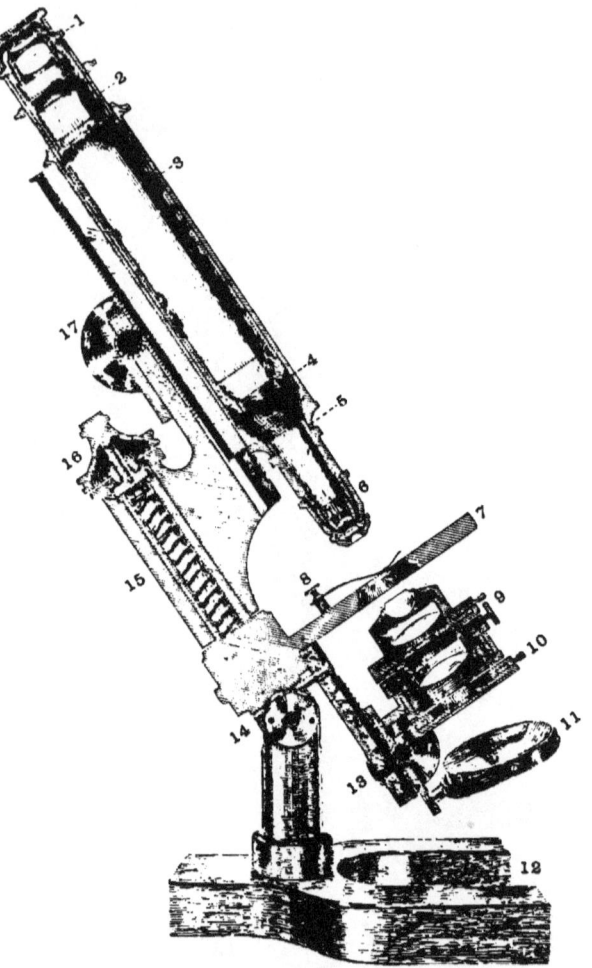

THE MICROSCOPE IN SECTION.

1. Compensation ocular 12; it is a positive ocular.
2. Draw-tube, by which the tube is lengthened or shortened.
3. Main tube or body, to the lower end of which the objective or revolving nose-piece is attached.
4. Society screw in the lower end of the draw-tube.
5. Society screw in the lower end of the tube.
6. Objective in position.
7. Stage, under which is the substage with the substage condenser.
8. Spring clip for holding the specimen.
9. Screw for centering, and handle of the iris diaphragm in the achromatic condenser (see Fig. 41).
10. Iris diaphragm outside the principal focus of the condenser for use in centering (§ 77).
11. Mirror with plane and concave faces.
12. Horse-shoe base.
13. Rack and pinion for the substage condenser.
14. Flexible pillar.
15. Part of pillar with spiral spring of fine adjustment.
16. Screw of fine adjustment.
17. Milled head of coarse adjustment.

THE MICROSCOPE

AND

MICROSCOPICAL METHODS.

CHAPTER I.

THE MICROSCOPE AND ITS PARTS.

APPARATUS AND MATERIAL FOR THIS CHAPTER.

A simple microscope (§ 2, 9); A compound microscope with nose-piece (Figs. 68-80), eye-shade (Figs. 59-60), achromatic (§ 18), apochromatic (§ 20), dry (§ 15), immersion (§ 16), unadjustable and adjustable objectives (§ 21, 22), Huygenian or negative (§ 35), positive (§ 34) and compensation oculars (§ 36), stage micrometer, homogeneous immersion liquid (§ 16, Ch. IV), benzin and distilled water (§ 103-108). Mounted letters or figures (§ 49); ground-glass and lens paper (§ 49).

A MICROSCOPE.

§ 1. A Microscope is an optical apparatus with which one may obtain a clear image of a near object, the image being always larger than the object; that is, it enables the eye to see an object under a greatly increased visual angle, as if the object were brought very close to the eye without affecting the distinctness of vision. Whenever the microscope is used for observation, the eye of the observer forms an integral part of the optical combination (Figs. 16, 21).

§ 2. A Simple Microscope.—With this an enlarged, erect image of an object may be seen. It always consists of one or more converging lenses or lens-systems (Figs. 16-20), and the object must be placed within the principal focus (§ 9). The simple microscope may be held in the hand or it may be mounted in some way to facilitate its use (Figs. 17-20).

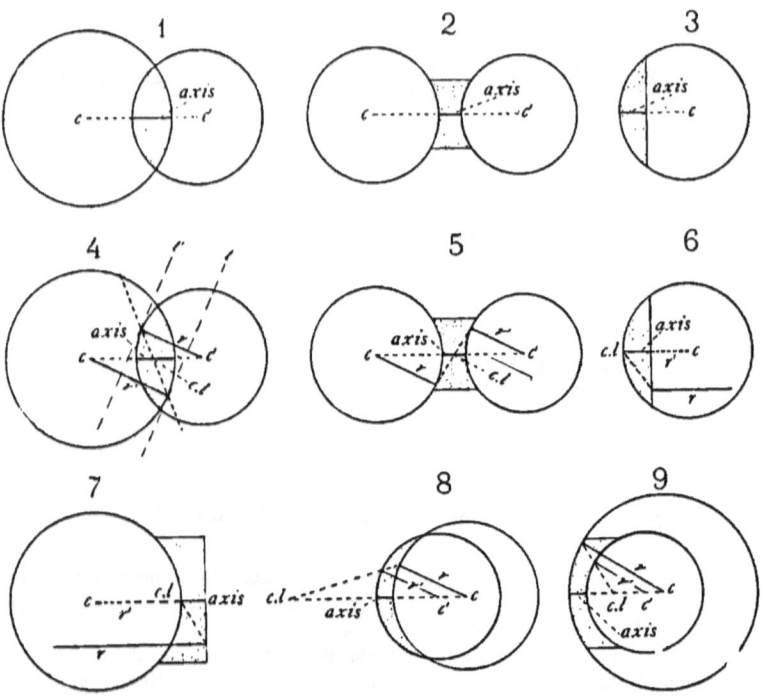

FIGS. 1-9, *showing the Principal Optic Axis and the Optical Center of various forms of Lenses.*

Axis. *The Principal Optic Axis.* c-c'. *Centers of curvature of the two surfaces of the lens.* c. l. *Optical center of the lens.* r-r'. *Radii of curvature of the two lens surfaces.* t-t'. *Tangents in Fig. 4.*

§ 3. **Principal Optic Axis.**—In spherical lenses, *i.e.*, lenses whose surfaces are spherical, the Axis is the part of the line joining the centers of curvature and traversing the lens; it is the unbroken part of the line c-c' in all the figures. In lenses with one plane surface (Figs. 3, 6, 7) the radius of the plane surface is any line at right angles to it, but in determining the axis it must be the one which is continuous with the radius of the curved surface, consequently the axis in such lenses is on the radius of the curved surface which meets the plane surface at right angles.

§ 4. **Optical Center.**—The optical center of a lens is the point through which rays pass without angular deviation, that is, the emergent ray is parallel to the incident ray. It is determined geometrically by drawing parallel radii of the curved surfaces, r-r' in Figs. 4-9, and joining the peripheral ends of the radii. The optical center is the point on the axis cut by the line joining the radii. In Figs. 4-5 it is within the lens; in 6-7 it is at the curved surface, and in the meniscus 8, 9) it is wholly outside the lens, being situated on the side of the greater curvature.

In determining the center in a lens with a plane surface, the conditions can be satisfied only by using the radius of the curved surface which is continuous with the axis of the lens, then any line at right angles to the plane surface will be parallel with it, and may be considered part of the radius of the plane surface. (That is, a plane surface may be considered part of a sphere with infinite radius, hence any line meeting the plane surface at right angles may be considered as the peripheral part of the radius.) In Figs. 6, 7, (r') is the radius of the curved surface and (r) of the plane surface; and the point where a line joining the ends of these radii crosses the axis is at the curved surface in each case.

By a study of Fig. 4 it will be seen that if tangents be drawn at the peripheral ends of the parallel radii, the tangents will also be parallel and a ray incident at one tangential point and traversing the lens and emerging at the other tangential point acts as if traversing, and is practically traversing a piece of glass which has parallel sides at the point of incidence and emergence, therefore the emergent ray will be parallel with the incident ray. This is true of all rays traversing the center of the lens.

§ 5. Secondary Axis.—Every ray traversing the center of the lens, except the principal axis, is a secondary axis; and every secondary axis is more or less oblique to the principal axis. In Fig. 14, line (2), is a secondary axis, and in Fig. 15, line (1). See also Fig. 57.

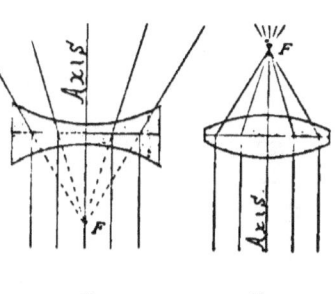

FIGS. 10, 11.—*Sectional views of a concave or diverging and a convex or converging lens to show that in the concave lens the principal focus is virtual as indicated by the dotted lines, while with the convex lens the focus is real and on the side of the lens opposite to that from which the light comes. The principal focal distance is the distance along the axis, from the optical center to the principal focus (F).*

§ 6. Principal Focus.—This is the point where parallel rays traversing the lens cross the axis; and the distance from the focus to the center of the lens measured along the axis is the *Principal Focal Distance.* In the diagrams, Fig. 10 is seen to be a diverging lens and the rays cross the axis only by being projected backward. Such a focus is said to be virtual, as it it has no real existence. In Fig. 11 the rays do cross the axis and the focus is said to be real. If the light came from the opposite direction it would be seen that there is a principal focus on the other side, that is there are two principal foci, one on each side of the lens. These two foci are both principal foci; and as there may be foci on secondary axes also, each focus on a secondary axis has its conjugate. In the formation of images the image is the conjugate of the object and conversely the object is the conjugate of the image.

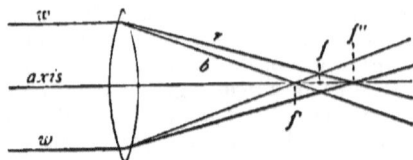

FIG. 12.—*Double Convex Lens, Showing Chromatic Aberration.*

The ray of white light (w) is represented as dividing into the short waved, blue (b) and the long waved, red (r) light. The blue (b) ray comes to a focus nearer the lens and the red ray (r) farther from the lens than the principal focus (f). Principal focus (f) for rays very near the axis, f' and f'', foci of blue and red light coming from near the edge of the lens. The intermediate wave lengths would have foci all the way between f' and f''.

§ 7. **Chromatic Aberration.**—This is due to the fact that ordinary light consists of waves of varying length, and as the effect of a lens is to change the direction of the waves, it changes the direction of the short waves more markedly than the long waves. Therefore the short waved, blue light will cross the axis sooner than the long waved, red light, and there will result a superposition of colored images, none of which are perfectly distinct. (Fig. 12).

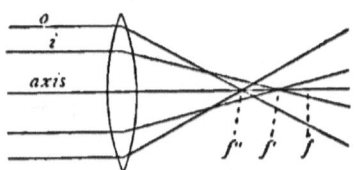

FIG. 13. *Double Convex Lens, showing Spherical Aberration.*

FIG. 13. *The ray (o) near the edge of the lens is brought to a focus nearer the lens than the ray (i). Both are brought to a focus sooner than rays very near the axis. (f) Principal focus for rays very near the axis; (f') Focus for the ray (i), and (f'') Focus for the ray (o). Intermediate rays would cross the axis all the way from (f'' to f).*

§ 8. **Spherical Aberration.**—This is due to the unequal turning of the light in different zones of a lens. The edge of the lens refracts proportionally too much and hence the light will cross the axis or come to a focus nearer the lens than a ray which is nearer the middle of the lens. Thus, in Fig. 13, if the focus of parallel rays very near the axis is at f, rays ($o\ i$) nearer the edge would come to a focus nearer the lens, the focus of the ray nearest the edge being nearest the lens. Every simple lens has the defect of both chromatic and spherical aberration, and to overcome this, kinds of glass of different refractive power and different dispersive power are combined, concave lenses neutralizing the defects of convex lenses. If the concave lens is not sufficiently strong to neutralize the aberration of the convex lens, the combination is said to be *under-corrected*, while if it is too strong and brings the marginal rays or the blue rays to a focus beyond the true principal focus, the combination is *over-corrected*.

Probably no higher technical skill is used in any art than is requisite in the preparation of microscopical objectives, oculars and illuminators.

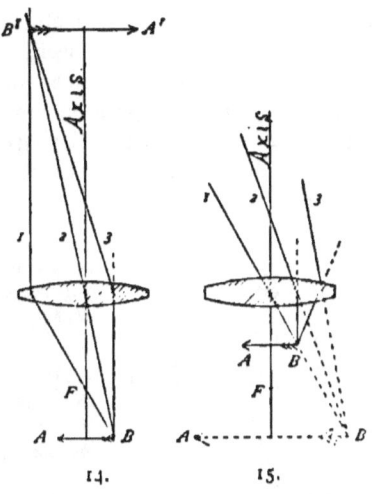

FIGS. 14 AND 15. 14. *Convex lens showing the position of the object (A–B) outside the principal focus (F), and the course of the rays in the formation of real images. To avoid confusion the rays are drawn from only one point.*
A B. Object outside the principal focus. B' A'. Real, enlarged image on the opposite side of the lens.
Axis. Principal optic axis. 1, 2, 3. Rays after traversing the lens. They are converging, and consequently form a real image. The dotted line and the line (2) give the direction of the rays as if unaffected by the lens. (F). The principal focus.

FIG. 15. *Convex lens showing the position of the object (A B) within the principal focus and the course of rays in the formation of a virtual image.*
A B. The object placed between the lens and its focus; A' B' virtual image formed by tracing the rays backward. It appears on the same side of the lens as the object, and is erect (§ 9).
Axis. The principal optic axis of the lens. F. The principal focus.
1, 2, 3. Rays from the point B of the object. They are diverging after traversing the lens, but not so divergent as if no lens were present, as is shown by the dotted lines. Ray (1) traverses the center of the lens, and is therefore not deviated. It is a secondary axis (§ 5).

§ 8a. **Geometrical Construction of Images.**—As shown in Figs. 14-15, for the determination of any point of an image, or the image being known, to determine the corresponding part of the object, it is necessary to know the position of the principal focus (and there is one on each side of the lens, § 6), and the optical center (Figs. 1-9) of the lens. Then a secondary axis, (2) in Fig. 14, (1) in Fig. 15, is drawn from the extremity of the object and prolonged indefinitely above the lens, or below it for virtual images. A second line is drawn from the extremity of the object, (3) in Fig. 14, (2) in Fig. 15, to the lens parallel with the principal axis. After traversing the lens it must be drawn through the principal focal point.

If now it is prolonged it will cross the secondary axis above the lens for a real image and below for a virtual image. The crossing point of these lines determines the position of the corresponding part of the image. Commencing with any point of the object the corresponding point of the image may be determined as just described, and conversely commencing with the image corresponding points of the object may be determined.

SIMPLE MICROSCOPE: EXPERIMENTS.

§ 9. Employ a tripod or other simple microscope, and for object a printed page. Hold the eye about two centimeters from the upper surface of the magnifier, then alternately raise and lower the magnifier until a clear image may be seen. (This mutual arrangement of microscope and object so that a clear image may be seen, is called focusing). When a clear image is seen, note that the letters appear as with the unaided eye except that they are larger, and the letters appear erect or right side up, instead of being inverted, as with the compound microscope (§ 10, 49).

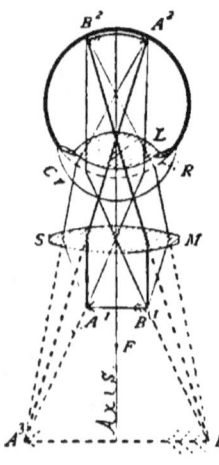

FIG. 16. *Diagram of the simple microscope showing the course of the rays and all the images, and that the eye forms an integral part of it.*

$A^1 B^1$. *The object within the principal focus.* $A^2 B^2$. *The virtual image on the same side of the lens as the object. It is indicated with dotted lines, as it has no actual existence.*

$B^3 A^3$. *Retinal image of the object ($A^1 B^1$). The virtual image is simply a projection of the retinal image in the field of vision.*

Axis. *The principal optic axis of the microscope and of the eye.* Cr. *Cornea of the eye.* L. *Crystalline lens of the eye.* R. *Ideal refracting surface at which all the refractions of the eye may be assumed to take place.*

Hold the simple microscope directly toward the sun and move it away from and toward a piece of printed paper until the smallest bright point on the paper is obtained. This is the *burning point* or *focus*, and as the rays of the sun are nearly parallel, the burning point represents approximately the principal focus (Figs. 10, 11). Without changing the position of the paper or the magnifier, look into the magnifier and note that the letters are very indistinct or invisible. Move the magnifier a centimeter or two farther from the paper and no image can be seen. Now move the magnifier closer to the paper, that is, so that it is less than the focal distance from the paper, and the letters will appear distinct. This shows that in order to see a distinct image with a simple microscope, the object must always be nearer to it than its principal focal point. Or, in other words, the object must be within the principal focus. Compare (§ 49.).

FIG. 17.
Tripod Magnifier.

After getting as clear an image as possible with a simple microscope, do not change the position of the microscope but move the eye nearer and farther from it, and note that when the eye is in one position, the largest field may be seen. This position corresponds to the eye-point (Fig. 30) of an ocular, and is the point at which the largest number of rays from the microscope enter the eye. Note that the image appears on the same side of the magnifier as the object.

Simple microscopes are very convenient when only a small magnification (Ch. IV) is desired, as for dissecting. Achromatic triplets are excellent and convenient for

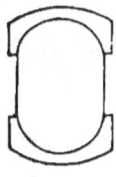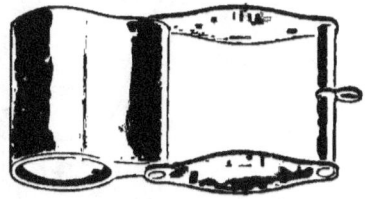

FIG. 18. *Achromatic Triplet for the pocket. As shown in the left hand figure it is composed of three lenses, one of crown and two of flint glass. The whole is protected by a metal covering when not in use. (Bausch & Lomb Opt. Co.)*

the pocket (Fig. 18). For use in conjunction with a compound microscope, the tripod magnifier (Fig. 17) is one of the best forms. For many purposes a special mechanical mounting like that of Figs. 19, 20, is to be preferred.

COMPOUND MICROSCOPE.

§ 10. **A Compound Microscope.**—This enables one to see an enlarged, inverted image. It always consists of two optical parts—an *objective*, to produce an enlarged, inverted, real image of the object, and an *ocular* acting in general like a simple microscope to magnify this real image (Fig. 21). There is also usually present a mirror, or both a mirror and some form of condenser or illuminator for lighting the object. The stand of the microscope consists of certain mechanical arrangements for holding the optical parts and for the more satisfactory use of them. (See frontispiece).

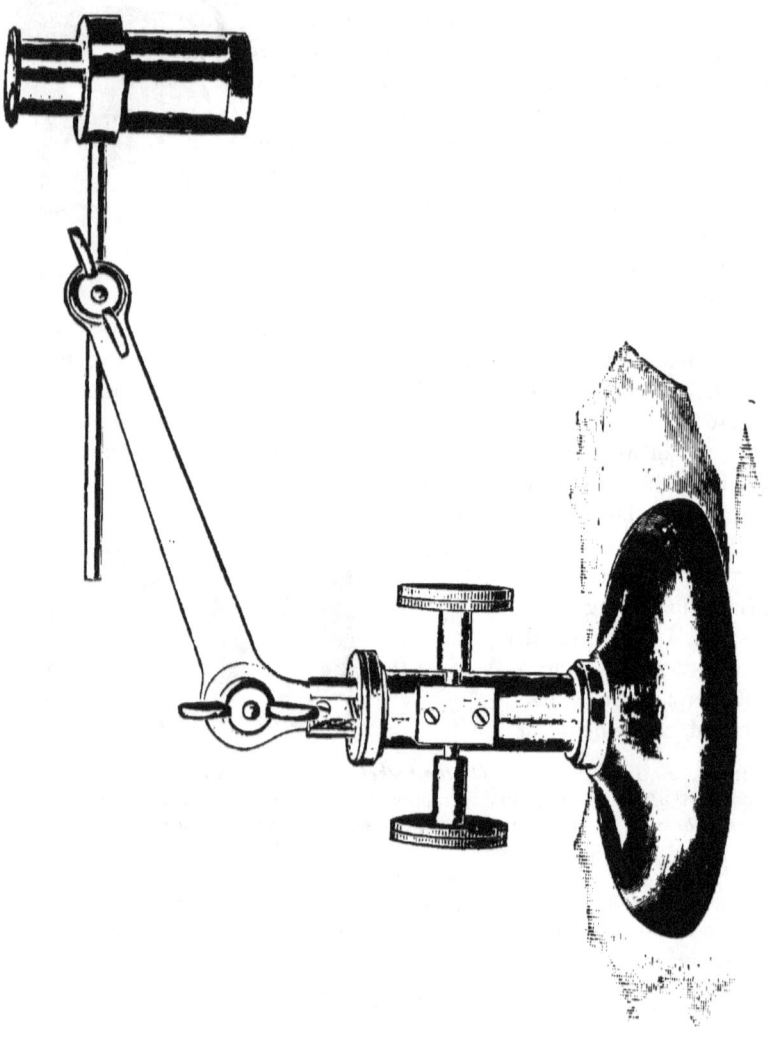

FIG. 19. *Lens Holder with adjustments for focusing and for turning the lens in any direction. This is especially useful in dissecting the minute parts of animals too large for the regular dissecting microscope (Fig. 20). (Bausch & Lomb Opt. Co.).*

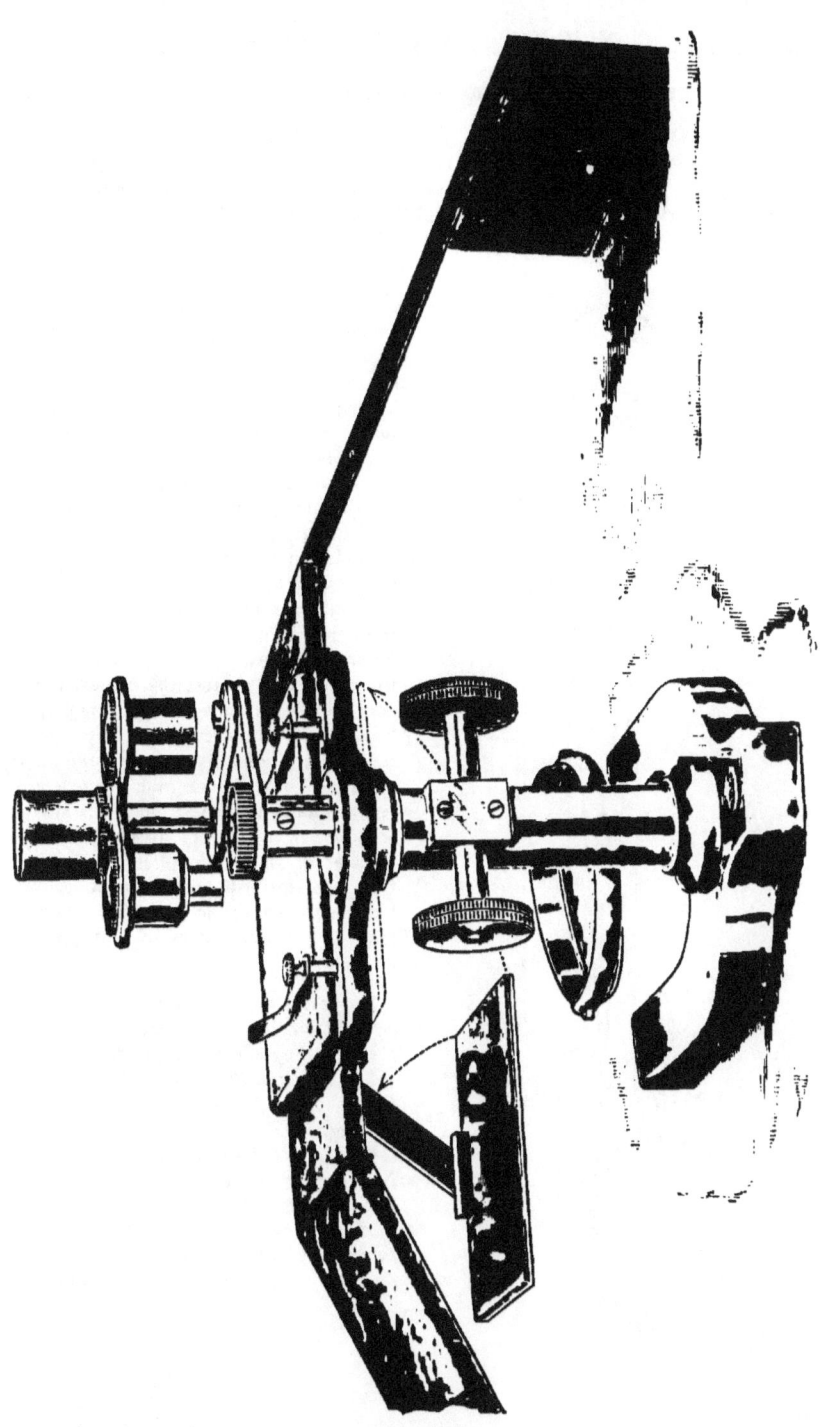

FIG. 20. *Dissecting Microscope with hand rests and nose-piece for several lenses of different power.* (*Bausch & Lomb Optical Co.*).

MECHANICAL PARTS.

§ 11. The Mechanical Parts of a laboratory, compound microscope are shown in the frontispiece, and are described in the explanation of that figure. The student should study the figure with a microscope before him and become thoroughly familiar with the names of all the parts. See also the cuts of microscopes at the end of Ch. II.

OPTICAL PARTS.

§ 12. Microscopic Objective.—This consists of a converging lens or of one or more converging lens-systems, which give an enlarged, inverted, real image of the object (Figs. 14, 21). And as for the formation of real images in all cases, the object must be placed outside the principal focus, instead of within it, as for the simple microscope. (See §§ 9, 49, Figs. 16, 21).

Modern microscopic objectives usually consist of two or more systems or combinations of lenses, the one next the object being called the *front combination* or lens, the one farthest from the object and nearest the ocular, the *back combination* or system. There may be also one or more intermediate systems. Each combination is, in general, composed of a convex and a concave lens. The combined action of the systems serves to produce an image free from color and from spherical distortion. In the ordinary achromatic objectives the convex lenses are of crown and the concave lenses of flint glass (Figs. 22, 23).

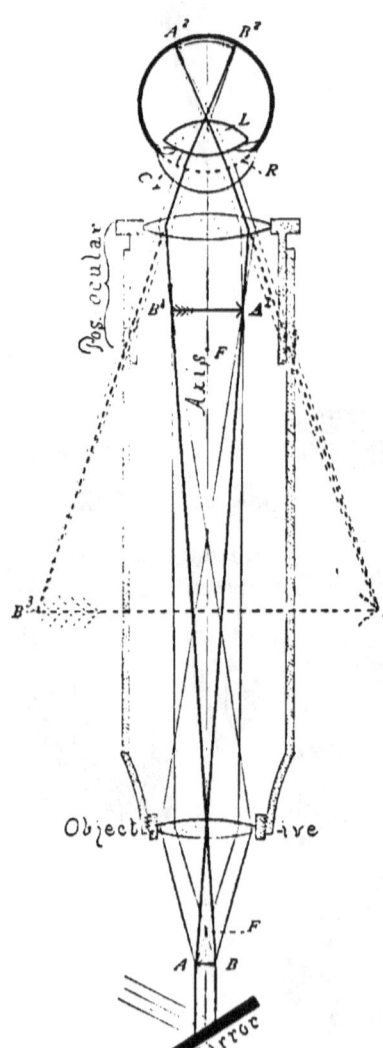

FIG. 21. *Diagram showing the principle of a compound microscope with the course of the rays from the object (A B) through the objective to the real image (B′ A′), thence through the ocular and into the eye to the retinal image (A² B²), and the projection of the retinal image into the field of vision as the virtual image (B³ A³).*

A B. *The object.* A² B². *The retinal image of the inverted real image, (B¹ A¹), formed by the objective.* B³ A³. *The inverted virtual image, a projection of the retinal image.*

Axis. The principal optic axis of the microscope and of the eye.
Cr. Cornea of the eye. *L.* Crystalline lens of the eye. *R.* Single, ideal, refracting surface at which all the refractions of the eye may be assumed to take place.
F.F. The principal focus of the positive ocular and of the objective.
Mirror. The mirror reflecting parallel rays to the object. The light is central. See Ch. II.
Pos. Ocular. An ocular in which the real image is formed outside the ocular. Compare the positive ocular with the simple microscope (Fig. 16).

NOMENCLATURE OR TERMINOLOGY OF OBJECTIVES.

§ 13 **Equivalent Focus.**—In America, England, and sometimes also on the Continent, objectives are designated by their equivalent focal length. This length is given either in inches (usually contracted to in.) or in millimeters (mm). Thus: An objective designated $\frac{1}{12}$ in. or 2 mm., indicates that the objective produces a real image of the same size as is produced by a simple converging lens whose principal focal distance is $\frac{1}{12}$ inch or 2 millimeters (Fig. 11). An objective marked 3 in. or 75 mm., produces approximately the same sized real image as a simple converging lens of 3 inches or 75 millimeters focal length. And in accordance with the law *that the relative size of object and image vary directly as their distance from the center of the lens* (Figs 14, 15, see Ch. IV,) it follows that the less the focal distance of the simple lens or of the equivalent focal distance of the objective, the greater is the size of the real image, as the tube-length remains constant and the image in all cases is found at about 160 or 250 mm. from the objective.

§ 14 **Numbering or Lettering Objectives.**—Instead of designating objectives by their equivalent focus, many Continental opticians use letters or figures for this purpose. With this method the smaller the number, or the earlier in the alphabet the letter, the lower is the power of the objective. (See further in Ch. IV, for the power or magnification of objectives). This method is entirely arbitrary and does not, like the one above, give direct information concerning the objective.

§ 15. **Dry Objectives.**—These are objectives in which the space between the front of the objective and the object or cover-glass is filled with air (Fig. 22). Most objectives of low and medium power (*i. e.*, $\frac{1}{5}$th in. or 3 mm. and lower powers) are dry.

FIG. 22. *Section of a dry objective showing working distance and lighting by reflected light.*
Axis. The principal optic axis of the objective.
B C. Back Combination, composed of a plano-concave lens of flint glass (*F*), and a double convex lens of crown glass (*c*).
F C. Front Combination.
C, O, sl. The cover-glass, object and slide.
Mirror. The mirror is represented as above the stage, and as reflecting parallel rays from its plane face upon the object.
Stage. Section of the stage of the microscope.
W. The Working Distance, that is the distance from the front of the objective to the object when the objective is in focus.

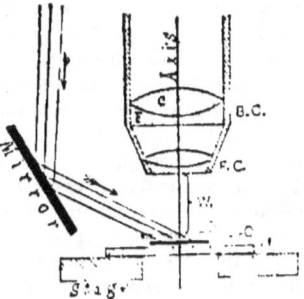

§ 16. **Immersion Objectives.**—An immersion objective is one with which there is some liquid placed between the front of the objective and the object or cover-glass. The most common immersion objectives are those (A) in which water is used as the immersion fluid, and (B) where some liquid is used having the same refractive and dispersive power as the front lens of the objective. Such a liquid is called homogeneous, as it is optically homogeneous with the front glass of the objective. It may consist of thickened cedar wood oil or of glycerin containing some salt, as stannous chlorid, in solution. When oil is used as the immersion fluid the objectives are frequently called oil immersion objectives. The disturbing effect of the cover-glass (Fig. 56) is almost wholly eliminated by the use of homogeneous immersion objectives, as the rays undergo very little or no refraction on passing from the cover-glass through the immersion medium and into the objective; and when the object is mounted in balsam there is practically no refraction in the ray from the time it leaves the balsam till it enters the objective.

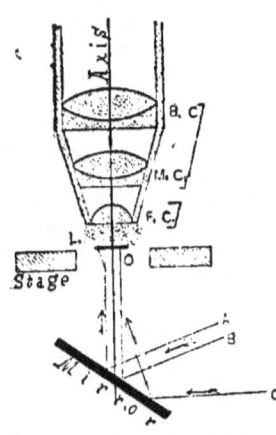

FIG. 23. *Sectional view of an Immersion, Adjustable Objective, and the object lighted with axial or central and with oblique light.*

Axis. The principal optic axis of the objective.
B C, M C, F C. The back, middle and front combination of the objective. In this case the front is not a combination, but a single plano convex-lens.
A, B. Parallel rays reflected by the mirror axially or centrally upon the object.
C. Ray reflected to the object obliquely.
I. Immersion fluid between the front of the objective and the cover-glass or object (O).
Mirror. The mirror of the microscope.
O. Object. It is represented without a cover-glass. Ordinarily objects are covered whether examined with immersion or with dry objectives.
Stage. Section of the stage of the microscope.

§ 17. **Non-Achromatic Objectives.**—These are objectives in which the chromatic aberration is not corrected, and the image produced is bordered by colored fringes. They show also spherical aberration and are used only on very cheap microscopes. (§§ 7, 8, Figs. 12, 13).

§ 18. **Achromatic Objectives.**—In these the chromatic and the spherical aberration are both largely eliminated by combining concave and convex lenses of different kinds of glass "so disposed that their opposite aberrations shall correct each other." All the better forms of objectives are achromatic and also aplanatic (§ 19). That is the various spectral colors come to the same focus.

§ 19. **Aplanatic Objectives, etc.**—These are objectives or other pieces of optical apparatus (oculars, illuminators, etc.), in which the spherical distortion is wholly or nearly eliminated, and the curvatures are so made that the central and marginal parts of the objective focus rays at the same point or level. Such pieces of apparatus are usually achromatic also. (§ 7, 8).

§ 20. **Apochromatic Objectives.**—A term used by Abbe to designate a form of objective made by combining new kinds of glass with a natural mineral (Calcium fluoride, Fluorite, or Fluor spar). The name, Apochromatic, is used to indicate the higher kind of achromatism in which rays of three spectral colors are combined at one focus, instead of rays of two colors, as in the ordinary achromatic objectives. At the present time (1896) several opticians make apochromatic objectives without using the fluorite. Some of the apochromatics deteriorate rather quickly in hot, moist climates.

The special characteristics of these objectives, when used with the "compensating oculars" are as follows:

(1) *Three rays* of different color are brought to one focus, leaving a small tertiary spectrum only, while with objectives as formerly made from crown and flint glass, only *two* different colors could be brought to the same focus.

(2) In these objectives the correction of the spherical aberration is obtained for *two* different colors in the brightest part of the spectrum, and the objective shows the same degree of chromatic correction for the marginal as for the central part of the aperture. In the old objectives, correction of the spherical aberration was confined to rays of *one* color, the correction being made for the central part of the spectrum, the objective remaining *under* corrected spherically for the red rays and *over*-corrected for the blue rays (§ 8).

(3) The optical and chemical foci are identical, and the image formed by the chemical rays is much more perfect than with the old objectives, hence the new objectives are well adapted to photography.

(4) These objectives admit of the use of very high oculars, and seem to be a considerable improvement over those made in the old way with crown and flint glass. According to Dippel (Z. w. M. 1886, p. 300) dry apochromatic objectives give as clear images as the same power water immersion objectives of the old form.

§ 21. **Non-Adjustable or Unadjustable Objectives.**—Objectives in which the lenses or lens systems are permanently fixed in their mounting so that their relative position always remains the same. Low power objectives and those with homogeneous immersion are mostly non-adjustable. For beginners and those unskilled in manipulating adjustable (§ 22) objectives, non-adjustable ones are more satisfactory, as the optician has put the lenses in such a position that the most satisfactory results may be obtained when the proper thickness of cover-glass and tube-length are employed. (See table of tube-length and thickness of cover-glass below Fig. 24).

§ 22. **Adjustable Objectives.**—An adjustable objective is one in which the distance between the systems of lenses (usually the front and the back systems) may be changed by the observer at pleasure. The object of this adjustment is to correct or compensate for the displacement of the rays of light produced by the mounting medium and the cover-glass after the rays have left the object. It is also to compensate for variations in "tube-length." See § 24. As the displacement of the rays by the cover-glass is the most constant and important, these objectives are usually designated as having cover-glass adjustment or correction. (Fig. 23. See also practical work with adjustable objectives § 96).

§ 23. **Parachromatic, Panachromatic and Semi-apochromatic Objectives.**—These are trade names for objectives, most of them containing one or more lenses of the new Jena glass. They are said to approximate much more closely to the apochromatics than to the ordinary objectives.

§ 24. **Variable Objective.**—This is a low power objective of 36 to 26 mm. equivalent focus, depending upon the position of the combinations. By means of a screw collar the combinations may be separated, diminishing the power or approximated and thereby increasing it.

§ 25. **Projection Objectives.**—These are designed especially for projecting an image on a screen and for photo-micrography. They are characterized by having a flat, sharp field brilliantly lighted. In power they vary, the lowest being of 75 mm. and the highest of 6 mm. equivalent focus (see Ch. IV).

§ 26. **Illuminating or Vertical Illuminating Objectives.**—These are designed for the study of opaque objects with good reflecting surfaces, like the rulings on metal bars. The light enters the side of the tube or objective and is reflected vertically downward through the objective and thereby is concentrated upon the object. The object reflects part of the light back into the microscope thus enabling one to see a clear image.

§ 27. **Tube-Length and Thickness of Cover-Glasses.**—"In the construction of microscopic objectives, the corrections must be made for the formation of the image at a definite distance, or in other words the tube of the microscope on which the objective is to be used must have a definite length. Consequently the microscopist must know and use this distance or 'microscopical tube-length' to obtain the best results in using any objective in practical work." Unfortunately different opticians have selected different tube-lengths and also different points between which the distance is measured, so that one must know what is meant by the tube-length of each optician whose objectives are used. See table.

The thickness of cover-glass used on an object (see Ch. VII, on mounting), except with homogeneous immersion objectives, has a marked effect on the light passing from the object (Fig. 56). To compensate for this the relative positions of the systems composing the objective are different from what they would be if the object were uncovered. Consequently, in non-adjustable objectives some standard thickness of cover-glass is chosen by each optician and the position of the systems arranged accordingly. With such an objective the image of an uncovered object would be less distinct than a covered one, and the same result would follow the use of a cover-glass much too thick.

Length in Millimeters and Parts included in "Tube-Length" by Various Opticians.*

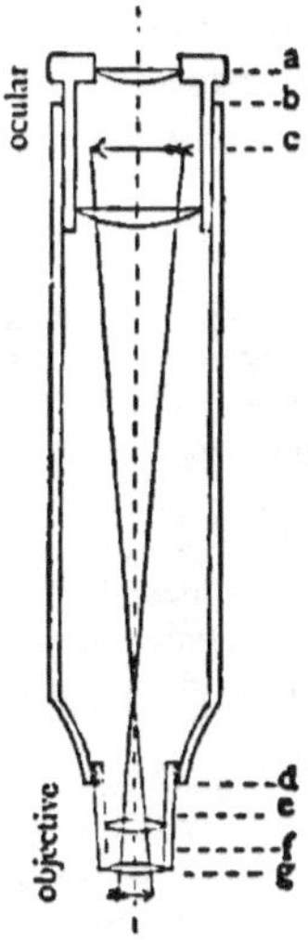

Pts. included in "Tube-length."
a
b See Diagram.
c

Optician	"Tube-length" in Millimeters
Grunow, New York	203 mm.
E. Leitz, Wetzlar	170 mm.
Nachet et Fils, Paris	146 or 200 mm.
Powell and Lealand, London	254 mm.
C. Reichert, Vienna	160 to 180 mm.
Spencer Lens Co., Buffalo	235 or 160 mm.
W. Wales, New York	254 mm.
Bausch & Lomb Opt. Co., Rochester	216 or 160 mm.
Bézu, Hausser et Cie, Paris	220 mm.
Klönne und Müller, Berlin	160–180 or 254 mm.
W. & H. Seibert, Wetzlar	190 mm.
Swift & Son, London	165 to 228½ mm.
C. Zeiss, Jena	160 or 250 mm.
Gundlach Optical Co., Rochester	254 mm.
R. Winkel, Göttingen	220 mm.
Ross & Co., London	254 mm.
R. & J. Beck, London	254 mm.
J. Green, Brooklyn	254 mm.
Hartnack, Potsdam	160–180 mm.
Vérick (Stiassnie) Paris	160–200 mm.
Watson & Sons, London	160–250 mm.

a-d { Powell and Lealand, ... W. Wales }
b-d { Bausch ... C. Zeiss }
a-g . Gundlach
a-g . R. Winkel
c-d . Ross
c-e . R. & J. Beck
e-f . J. Green
(Hartnack, Vérick, Watson)

Fig. 24.

Thickness of Cover Glass for Which Non-Adjustable Objectives are Corrected by Various Opticians.

Thickness	Optician
$\frac{25}{100}$ mm.	J. Green, Brooklyn. J. Grunow, Brooklyn. Powell and Lealand, London. Spencer Lens Co., Buffalo. W. Wales, New York.
$\frac{20}{100}$ mm. $\frac{18}{100}$ mm.	Watson & Sons, London. Klönne und Müller, Berlin.
$\frac{17}{100}$ mm.	E. Leitz, Wetzlar. R. Winkel, Göttingen, Germany.
$\frac{16}{100}$–$\frac{18}{100}$ mm.	Ross & Co., London.
$\frac{16}{100}$ mm.	Bausch & Lomb Optical Co., Rochester.
$\frac{12}{100}$–$\frac{20}{100}$ mm.	C. Zeiss, Jena.
$\frac{10}{100}$–$\frac{18}{100}$ mm.	C. Reichert, Vienna.
$\frac{15}{100}$ mm.	Gundlach Optical Co., Rochester. W. & H. Seibert, Wetzlar. R. & J. Beck, London.
$\frac{12}{100}$–$\frac{17}{100}$ mm.	J. Zentmayer, Philadelphia.
$\frac{10}{100}$–$\frac{15}{100}$ mm.	Nachet et Fils, Paris. Bézu, Hausser et Cie, Paris.
$\frac{10}{100}$ mm.	Swift and Son, London.

*The information contained in these tables was very kindly furnished by the opticians named, or by consulting catalogs. In most of the later catalogs the information is definite, and many makers now not only put their names and the equivalent focal length on their objectives, but they add the numerical aperture (§ 29) and the tube-length for which the objective is corrected. This is in accordance with the recommendations of the author in the original paper on "tube-length," (Proc. Amer. Soc. Micr., Vol. IX, p. 168, also by Bausch, Vol. XII, p.

§ 28. **Aperture of Objectives.**—The angular aperture or angle of aperture of an objective is the angle "contained, in each case, between the most diverging of the rays issuing from the axial point of an object [*i. e.*, a point in the object situated on the extended optic axis of the microscope], that can enter the objective and take part in the formation of an image." (Carpenter).

43). If the table in this edition is compared with the original table or with that in the previous edition of this book some differences will be noted, the changes being in the direction of uniformity and in general in the direction recommended by the writer and Mr. Bausch and the committee of the American Microscopical Society. The recommendations of the committee, published in the Proceedings, Vol. XII, p. 250, are as follows:

"Believing in the desirability of a uniform tube-length for microscopes, we unanimously recommend: 1. That the parts of the microscope included in the tube-length should be the same by all opticians, and that the parts included should be those between the upper end of the tube where the ocular is inserted and the lower end of the tube where the objective is inserted.

Fig. 25. *The tube of a microscope with ocular micrometer and nose piece in position to show that in measuring tube-length one must measure from the eye lens to the place where the objective is attached.* (*Zeiss' Catalog, No. 30*).

2. That the actual extent of tube length as defined in section 1—Be, for the short or continental tube, 160 mm., or 6.3 inches, and 216 mm., or 8½ inches, for the long tube, and that the draw tube of the microscope possess two special marks indicating these standard lengths.

3. That oculars be made par-focal, and that the par-focal plane be coincident with that of the upper end of the tube.

4. That the mounting of all objectives of 6 mm. (¼ inch) and shorter focus should be such as to bring the optical center of the objective 1½ inches below the shoulder, and that all objectives be marked with the tube-length for which they are corrected.

5. That non-adjustable objectives be corrected for cover-glass from $\tfrac{15}{100}$ to $\tfrac{20}{100}$ mm. ($\tfrac{1}{160}$ to $\tfrac{1}{130}$ inch) in thickness.

These recommendations give a distance of 10 inches (254 mm.) between the par-focal plane of the ocular and the optical center of the objective for the long tube, and are essentially in accord with the actual practice of opticians.

At the request of the committee, a joint conference was held with the opticians belonging to the Society and present at the meeting. They expressed their belief in the entire practicability of the above recommendations and a willingness to adopt them."

(Signed)   SIMON H. GAGE,
                                                     A. CLIFFORD MERCER,
                                                     CHARLES E. BARR.

In general, the angle increases with the size of the lenses forming the objective and the shortness of the equivalent focal distance (¿ 13). If all objectives were dry or all water or all homogeneous immersion a comparison of the angular aperture would give one a good idea of the relative number of image forming rays.

FIG. 26. *Diagram illustrating the angular aperture of a microscopic objective. Only the front lens of the objective is shown.*

Axis, the principal optic axis of the objective.

B A, B C, the most divergent rays that can enter the objective, they mark the angular aperture. A B D or C B D half the angular aperture. This is designated by u in making Numerical Aperture computations. See the table, ¿ 30.

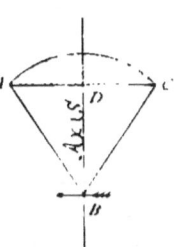

transmitted by different objectives; but as some are dry, others water and still others homogeneous immersion, one can see at a glance that, other things being equal, the dry objective (Fig. 27) receives less light than the water immersion, and the water immersion (Fig. 28) less than the homogeneous immersion (Fig. 29). In order to render comparison accurate between different kinds of objectives, Professor Abbe takes into consideration the rays actually passing from the back combination of the objective to form the real image; he thus takes into account the medium in front of the objective as well as the angular aperture. The term "*Numerical Aperture*," (N. A.) was introduced by Abbe to indicate the capacity of an optical instrument "for receiving rays from the object and transmitting them to the image."

¿ 29. **Numerical Aperture** (abbreviated N. A.), as now employed for microscope objectives, is the ratio of the semi-diameter of the emergent pencil to the focal length of the lens. Or as the factors are more readily obtainable it is simpler to utilize the relationship shown in the La Grange-Helmholtz formula, and indicate the aperture by the expression : N. A. $= n \sin u$. In this formula n is the index of refraction of the medium in front of the objective (air, water or homogeneous liquid), and $\sin u$ is the sine of half the angle of aperture Fig. 26, D B A). For the mathematical discussion showing that the expressions

$$\frac{\text{semi-diameter of emergent pencil}}{\text{focal length of the lens}} = n \sin u,$$ the student is referred to the *Journal of the Royal Microscopical Society*, 1881, pp. 392-395.

For example, take three objectives each of 3 mm. equivalent focus, one being a dry, one a water immersion, and one a homogeneous immersion. Suppose that the dry objective has an angular aperture of 106°, the water immersion of 94° and the homogeneous immersion of 90°. Simply compared as to their angular aperture, without regard to the medium in front of the objective, it would look as if the dry objective would actually take in and transmit a wider pencil of light than either of the others. However, if the medium in front of the objective is considered, that is to say, if the numerical instead of the angular apertures are compared, the results would be as follows: Numerical Aperture of a dry objective of 106°, N. A. $= n \sin u$. In the case of dry objectives the medium in front of the objective being air, the index of refraction is unity, whence $n = 1$. Half the angular aperture is $\frac{106°}{2} = 53°$. By consulting a table of natural sines it will be found that the sine of 53° is 0.799, whence N. A. $= n$ or $1 \times \sin u$ or $0.799 = 0.799$.

* ¿ 29a. **Interpolation.** In practice, as in solving problems similar to those on the following pages and those in refraction on p. 50, if one cannot find a sine exactly

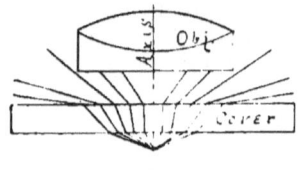

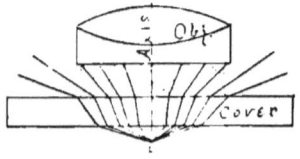

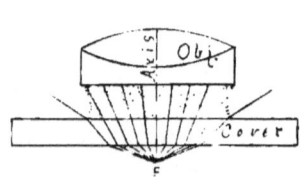

FIGS. 27–29 *are somewhat modified from Ellenberger, and are introduced to illustrate the relative amount of utilized light, with dry, water immersion and homogeneous immersion objectives of the same equivalent focus. The point from which the rays emanate is in air in each case. If Canada balsam were in place of the air beneath the cover glass there would be practically no refraction of the rays on entering the cover glass (¾ 16).*

FIG. 27. *Showing the course of the rays passing through a cover glass from an axial point of the object, and the number that finally enter the front of a dry objective.*

FIG. 28. *Rays from the axial point of the object traversing a cover of the same thickness as in Fig. 27, and entering the front lens of a water immersion objective.*

FIG. 29. *Rays from an axial point of the object traversing a cover glass and entering the front of a homogeneous immersion objective.*

With the water immersion objective the medium in front is water, and its index of refraction is 1.33, whence $n = 1.33$. Half the angular aperture is $\frac{94°}{2} = 47°$, and by the table the sine of $47°$ is corresponding to a given angle; or if one has an angle which does not correspond to any sine or angle given in the table, the sine or angle may be closely approximated by the method of interpolation, as follows: Find the sine in the table nearest the sine whose angle is to be determined. Get the difference of the sines of the angles greater and less than the sine whose angle is to be determined. That will give the increase of sine for that region of the arc for 15 minutes. Divide this increase by 15 and it will give with approximate accuracy the increase for 1 minute. Now get the difference between the sine whose angle is to be determined and the sine just below it in value. Divide this difference by the amount found necessary for an increase in angle of 1 minute and the quotient will give the number of minutes greater the sine is than the next lower one whose angle is known. Add this number of minutes to the angle of the next lower sine and the sum will represent the desired angle of the sine. Or if the sine whose angle is to be found is nearer in size to the sine just greater, proceed exactly as before, getting the difference in the sines, but subtract the number of minutes of difference and the result will give the angle sought. For example take the case in the last section where the sine of the angle of 28° 54′ is given as 0.48327. If one consults the table the nearest sines found are 0.48099, the sine of 28° 45′, and 0.48481, the sine of 29°. Evidently then the angle sought must lie between 28° 45′ and 29°. If the difference between 0.48481 and 0.48099 be obtained, 0.48481 − 0.48099 = 0.00382, and this increase for 15′ be divided by 15 it will give the increase for 1 minute; 0.00382÷15 = 0.000254. Now the difference between the sine whose angle is to be found and the next lower sine is 0.48327 − 0.48099 = 0.00228. If this difference is divided by the amount found necessary for 1 minute it will give the total minutes above 28° 45′; 0.00228 ÷ 0.000254 = 9. That is, the angle sought is 9 minutes greater than 28° 45′ = 28° 54′.

found to be 0.731, *i. e.*, sin u = 0.731, whence N. A. = n or 1.33 × sin u or 0.731 = 0.972.

With the oil immersion in the same way N. A. = n sin u ; n or the index of refraction of the homogeneous fluid in front of the objective is 1.52, and the semi-angle of aperture is $\frac{90°}{2}$ = 45°. The sine of 45° is 0.707, whence N. A. = n or 1.52 × sin u or 0.707 = 1.074.

By comparing these numerical apertures: Dry 0.799, water 0.972, homogeneous immersion 1.074, the same idea of the real light efficiency and image power of the different objectives is obtained, as in the graphic representations shown in Figs. 27-29.

If one knows the numerical aperture (N. A.) of an objective the angular aperture is readily determined from the formula ; and one can determine the equivalent angles of objectives used in different media (*i. e.*, dry or immersion). For example, suppose each of three objectives has a numerical aperture (N. A.) of 0.80, what is the angular aperture of each. Using the formula of N. A. = n sin u one has N. A. = 0.80 for all objectives.

For the dry objective n = 1 (Refractive index of air).
" water immersion objective n = 1.33 (Refractive index of water).
" homogeneous immersion objective n = 1.52 (Refractive index of homogeneous liquid). And 2 u is to be found in each case.

For the dry objective, substituting the known values the formula becomes 0.80 = 1 sin u, or sin u = 0.80. By inspecting the table of natural sines (3d page of cover) it will be found that 0.80 is the sine of 53 degrees and 8 minutes. As this is half the angle the entire angular aperture of the dry objective must be 53° 8' × 2 = 106° 16'.

For the water immersion objective, substituting the known values in the formula as before : 0.80 = 1.33 sin u, or sin u = $\frac{0.80}{1.33}$ = 0.6015. Consulting the table of sines as before, it will be found that 0.6015 is the sine of 36° 59' whence the angular aperture (water angle) is 36° 59' × 2 = 73° 58'.

For the homogeneous immersion objective, substituting the known values, the formula becomes : 0.80 = 1.52 sin u whence sin u = $\frac{0.80}{1.52}$ = 0.5263. And by consulting the table of sines it will be found that this is the sine of 31° 45½' whence 2 u or the entire angle (balsam or oil angle) is 63° 31'.

That is, three objectives of equal resolving powers, each with a numerical aperture of 0.80 would have an angular aperture of 106° 16' in air, 73° 58' in water, and 63° 31' in homogeneous immersion liquid.

For the apparatus and method of determining aperture, see appendix

§ 30. **Table of a Group of Objectives with the Numerical Aperture (N. A.)** *and the method of obtaining it. Half the angular aperture is designated by u and the index of refraction of the medium in front of the objective by n. For dry objectives this is air and $n = 1$, for water immersions $n = 1.33$, and for homogeneous immersions $n = 1.52$. (For a table of natural sines, see third page of cover).*

Objective.	Angular Aperture ($2u$)	Natural Sine of half the angular aperture. ($\sin u$).	Index of Refraction of the medium in front of the objective. (n).	Numerical Aperture (N.A.) $= n \sin u$.
25 mm. (Dry.)	20°	$\sin \frac{20}{2} = 0.1736$	$n = 1$	N.A. $= 1 \times 0.1736 = 0.173$
25 mm. (Dry.)	40°	$\sin \frac{40}{2} = 0.3420$	$n = 1$	N.A. $= 1 \times 0.3420 = 0.342$
12½ mm. (Dry.)	42°	$\sin \frac{42}{2} = 0.3583$	$n = 1$	N.A. $= 1 \times 0.3583 = 0.358$
12½ mm. (Dry.)	100°	$\sin \frac{100}{2} = 0.7660$	$n = 1$	N.A. $= 1 \times 0.7660 = 0.766$
6 mm. (Dry.)	75°	$\sin \frac{75}{2} = 0.6087$	$n = 1$	N.A. $= 1 \times 0.6087 = 0.608$
6 mm. (Dry.)	136°	$\sin \frac{136}{2} = 0.9272$	$n = 1$	N.A. $= 1 \times 0.9272 = 0.927$
3 mm. (Dry.)	115°	$\sin \frac{115}{2} = 0.8434$	$n = 1$	N.A. $= 1 \times 0.8434 = 0.843$
3 mm. (Dry.)	163°	$\sin \frac{163}{2} = 0.9890$	$n = 1$	N.A. $= 1 \times 0.9890 = 0.989$
2 mm. Water. Immersion.	96° 12'	$\sin \frac{96° 12'}{2} = 0.7443$	$n = 1.33$	N.A. $= 1.33 \times 0.7443 = 0.99$
2 mm. Homogeneous Immersion.	110° 38'	$\sin \frac{110° 38'}{2} = 0.8223$	$n = 1.52$	N.A. $= 1.52 \times 0.8223 = 1.25$
2 mm. Homogeneous Immersion.	134° 10'	$\sin \frac{134° 10'}{2} = 0.9210$	$n = 1.52$	N.A. $= 1.52 \times 0.9210 = 1.40$

§ 31. **Significance of Aperture.**—As to the real significance of aperture in microscopic objectives, it is now an accepted doctrine that— the corrections in spherical and chromatic aberration being the same— (1) Objectives vary directly as their numerical aperture in their ability to define or make clearly visible minute details (resolving power). For example an objective of 4 mm. equivalent focus and a numerical aperture of 0.50 N. A. would define or resolve only half as many lines to the millimeter or inch as a similar objective of 1.00 N. A. So also an objective of 2 mm. focus and 1.40 N. A. would resolve only twice as many lines to the millimeter as a 4 mm. objective of 0.70 N. A. Thus it is seen that defining power is not a result of magnification but of aperture, otherwise the 2 mm. objective would resolve far more than twice as many lines as the 4 mm. objective.

(2) The illuminating power of an objective of a given focus is found to vary directly as the square of the numerical aperture (N. A.)^2. Thus if two 4 mm. objectives of N. A. 0.20 and N. A. 0.40 were compared as to their illuminating power it would be found from the above that they would vary as $0.20^2 : 0.40^2 = 0.0400 : 0.1600$ or $1 : 4$. That is the objective of 0.20 N. A. would have but ¼th the illuminating power of the one of 0.40 N. A.

In considering illuminating power the equivalent focal length must also be considered. If the N. A. were the same in a 3 mm. and a 6 mm. objective their illuminating power would vary directly with the square of the foci. Thus the illuminating power of the 6 and the 3 mm. objectives would be as $6^2 : 3^2$ or 36 to 9 or $4 : 1$, that is 4 times as great in the 6 mm. as in the 3 mm. objective. As the magnification of an objective varies indirectly as the equivalent focus, it follows also that the illuminating power will vary indirectly as the square of the magnification of the objective. The magnification of the 6 mm. is 42 and of the 3 mm. 84, whence the illuminating power of the two objectives are as $42^2 : 84^2$ or $1764 : 7056$ or $1 : 4$. As the ratio is inverse in this case the result is the same as before, that is 4 times as great for the 6 mm. as for the 3 mm. objective.

(3) The penetrating power, that is the power to see more than one plane, is found to vary as the reciprocal of the numerical aperture $\frac{1}{N. A.}$ so that in an objective of a given focus the greater the aperture the less the penetrating power.

In comparing the penetrating power of objectives of different foci, the numerical aperture being the same, it is found that the penetrating power increases directly as the square of the focus. For example, two objectives of the same N. A., one of 4 mm. and the other of 2 mm. focus, the penetrating power would be as $4^2 : 2^2$ or $16 : 4$ or $4 : 1$. That is, the numerical aperture remaining the same, the greater the equivalent focus the greater the penetration.

To briefly summarize: The numerical aperture is concerned with resolution and the resolving power varies directly as the numerical aperture, thus if the N. A. is doubled, twice as many lines to the millimeter or inch can be resolved.

With illuminating and penetrating power the equivalent focus of the objective must be considered as well as the numerical aperture. With objectives of the same equivalent focus, to double the N. A. is to increase the illuminating power 4-fold but the penetrating power is halved.

The numerical aperture remaining constant, the illuminating and penetrating power vary directly as the square of the equivalent focus; thus a 4 mm. objective would give four times the illuminating and penetrating power of a 2 mm. objective.

Of course when equivalent focus and numerical aperture both differ the problem becomes more complex.

For a consideration of the aperture question, its history and significance, see J. D. Cox, Proc. Amer. Micr. Soc., 1884, pp. 5–39 ; Jour. Roy. Micr. Soc., 1881, pp. 303, 348, 365, 388 ; 1882, pp. 300, 460 ; 1883, p. 790 ; 1884, p. 20. Carpenter-Dallinger, Chapters II and V.

THE OCULAR.

§ 32 **A Microscopic Ocular or Eye-Piece** consists of one or more converging lenses or lens systems, the combined action of which is, like that of a simple microscope, to magnify the real image formed by the objective.

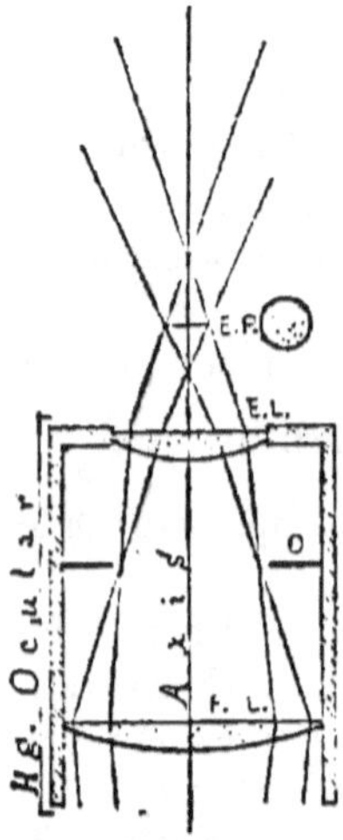

FIG. 30. *Sectional view of a Huygenian ocular (Hg. ocular), to show the formation of the Eye-Point.*
Axis. Optic axis of the ocular. *D. Diaphragm of the ocular. E. L. Eye-Lens. F. L. Field-Lens.*
E. P. Eye-point. As seen in section, it appears something like an hour-glass. *When seen as in looking into the ocular, i. e., in transection, it appears as a circle of light. It is at the point where the most rays cross.*

Depending upon the relation and action of the different lenses forming oculars, they are divided into two great groups, *negative* and *positive*.

§ 33. **Negative Oculars** are those in which the real, inverted image is formed within the ocular, the lower or field-lens serving to collect the image-forming rays somewhat, so that the real image is smaller than as if the field-lens were absent (Fig. 21). As the field-lens of the ocular aids in the formation of the real image it is considered by some to form a part of the objective rather than of the ocular. The upper or eye lens of the ocular magnifies the real image.

§ 34. **Positive Oculars** are those in which the real, inverted image of the objective is formed outside the ocular, and the entire system of ocular lenses magnifies the real image like a simple microscope (Fig. 16).

Positive and negative oculars may be readily distinguished, as a positive ocular may be used as a simple microscope, while a negative ocular cannot be so used when its field lens is in the natural position toward the object. By turning the eye-lens toward the object and looking into the field-lens an image may be seen, however.

Special names have also been applied to oculars, depending upon the designer, the construction, or the special use to which the ocular is to be applied. The following are most used.*

* In works and catalogs concerning the microscope and microscopic apparatus, and in articles upon the microscope in periodicals, various forms of oculars or eyepieces are so frequently mentioned, without explanation or definition, that it seemed worth while to give a list, with the French and German equivalents, and a brief statement of their character.

Achromatic Ocular; Fr. Oculaire achromatique; Ger. achromatisches Okular. Oculars in which chromatic aberration is wholly or nearly eliminated. *Aplanatic* Ocular; Fr. Oculaire aplanatique; Ger. aplanatisches Okular (see ¿ 19). *Binocular, stereoscopic* Ocular; Fr. Oculaire binoculaire stereoscopique; Ger. stereoskopisches Doppel-Okular. An ocular consisting of two oculars about as far apart as the two eyes. These are connected with a single tube which fits a monocular microscope. By an arrangement of prisms the image forming rays are divided, half being sent to each eye. The most satisfactory form was worked out by Tolles and is constructed on true stereotomic principles, both fields being equally illuminated. His ocular is also erecting. *Campani's* Ocular (see Huygenian Ocular). *Compound* Ocular; Fr. Oculaire composé; Ger. zusammengesetztes Okular. An ocular of two or more lenses, *e. g.*, the Huygenian (see Fig. 30). *Continental* Ocular. An ocular mounted in a tube of uniform diameter as in Fig. 31. *Deep* Ocular, see high ocular. *Erecting* Ocular; Fr. Oculaire redresseur; Ger. bildumkehrendes Okular. An ocular with which an erecting prism is connected so that the image is erect as with the simple microscope. Such oculars are most common on dissecting microscopes. *Filar micrometer* Ocular; Screw m. o. Cobweb m. o. Ger. Okular-Schraubenmikrometer. A modification of Ramsden's Telescopic Cobweb micrometer ocular. *Goniometer* Ocular; Fr. Oculaire à goniomètre; Ger. Goniometer-Okular. An ocular with goniometer for measuring the angles of minute crystals. *High* Ocular, sometimes called a deep ocular. One that magnifies the real image considerably, *i. e.*, 10 to 20 fold. *Huygenian* Ocular, Huygens' O., Campani's O., Airy's O.; Fr. Oculaire d'Huygens, o. de Campani; Ger. Huygens'sches Okular, Campaniches Okular, see ¿ 35. *Index* Ocular; Ger. Spitzen-O. An ocular with a minute pointer or two pointers at the level of the real image. The points are movable and serve for indicators and also, although not satisfactorily for micrometry. *Kellner's* Ocular, see orthoscopic ocular. *Low* Ocular, also called shallow ocular. An ocular which magnifies the real image only moderately, *i. e.*, 2 to 8 fold. *Micrometer* or *micrometric* Ocular; Fr. Oculaire micrometrique ou à micromètre; Ger. Mikrometer-Okular, Mess Okular, Bénèches O, Jackson m. o., see ¿ 38. *Microscopic* Ocular; Fr. Oculaire microscopic; Ger. mikroskopisches Okular. An ocular for the microscope instead of one for a telescope. *Negative* Ocular, see ¿ 33. Nelson's screw-micrometer ocular. A modification of the Ramsden's screw or cob-web micrometer in which positive compensating oculars may be used. *Orthoscopic* Oculars; also called Kellner's Ocular; Fr. Oculaire orthoscopique; Ger. Kellner'sches oder orthoskopisches Okular. An ocular with an eye-lens like one of the combinations of an objective (Figs. 22, 23) and a double convex field lens. The field-lens is in the focus of the eye-lens and there

§ 35. **Huygenian Ocular.**—A negative ocular designed by Huygens for the telescope, but adapted also to the microscope. It is the one now most commonly employed. It consists of a field-lens or collective (Fig. 30), aiding the objective in forming the real image, and an eye-lens which magnifies the real image. While

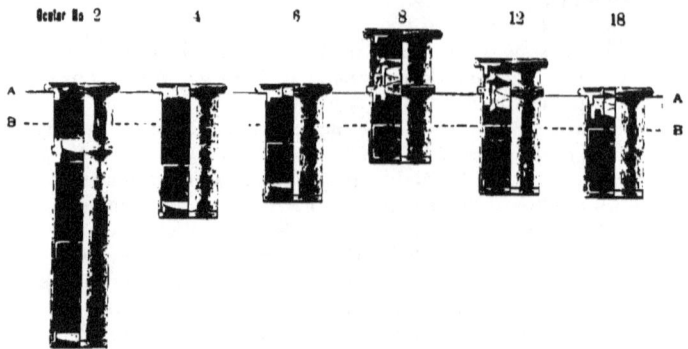

FIG. 31. *Compensating Oculars of Zeiss, with section removed to show the construction. The line A-A is at the level of the upper end of the tube of the microscope while B-B represents the lower focal points. It will be seen that the mounting is so arranged that the lower focal points in all are in the same plane and therefore the microscope remains in focus upon changing oculars. (The oculars are par-focal). The lower oculars, 2, 4 and 6 are negative, and the higher ones, 8, 12 18, are positive. The numbers 2, 4. 6, 8, 12, 18, indicate the magnification of the ocular. (From Zeiss' Catalog No. 30).*

is no diaphragm present. The field is large and flat. *Par-focal* Oculars, a series of oculars so arranged that the microscope remains in focus when the oculars are interchanged (Pennock, Micr. Bulletin, vol. iii, p. 9, 31). *Periscopic* Ocular; Fr. Oculaire periscopique; Ger. periskopisches Okular. A positive ocular devised by Gundlach. It consists of a double convex field-lens and a triplet eye-lens. It gives a large, flat field. *Positive* Ocular, see § 34. *Projection* Ocular; Fr. Oculaire de projection; Ger. Projections-Okular, see § 37. *Ramsden's* Ocular; Fr. Oculaire de Ramsden; Ger. Ramsden'sches Okular. A positive ocular devised by Ramsden. It consists of two plano-convex lenses placed close together with the convex surfaces facing each other. Only the central part of the field is clear. *Searching* Ocular; Fr. Oculaire d'orientation; Ger. Sucher-Okular, see § 36. *Shallow* Ocular, see low ocular. *Solid* Ocular, *holosteric* O.; Fr. Oculaire holostère; Ger. holosterisches Okular, Vollglass-Okular. A negative eye-piece devised by Tolles. It consists of a solid piece of glass with a moderate curvature at one end for a field-lens, and the other end with a much greater curvature for an eye-lens. For a diaphragm, a groove is cut at the proper level and filled with black pigment. It is especially excellent where a high ocular is desired. *Spectral* or *spectroscopic* Ocular; Fr. Oculaire spectroscopique; Ger. Spectral-Okular, see Microspectroscope, Ch. VI. *Stauroscopic* Ocular; Fr. Oculaire Stauroscopique. Ger. Stauroskop-Okular. An ocular with a Bertrand's quartz plate for mineralogical purposes. *Working* Ocular; Fr. Oculaire de travail; Ger. Arbeits Okular, see § 36.

the field-lens aids the objective in the formation of the real, inverted image, and increases the field of view. it also combines with the eye-lens in rendering the image achromatic.

§ 36. **Compensating Oculars.**—These are oculars specially constructed for use with the apochromatic objectives. They compensate for aberrations outside the axis which could not be so readily eliminated in the objective itself. An ocular of this kind, magnif ing but twice, is made for use with high powers, for the sake of the large field in finding objects; it is called a *searching ocular;* those ordinarily used for observation are in contradistinction called *working oculars.* Part of the compensating oculars are positive and part negative (Fig. 31).

§ 37. **Projection Oculars.**—These are oculars especially designed for projecting a microscopic image on the screen for class demonstrations, or for photographing with the microscope. While they are specially adapted for use with apochromatic objectives, they may also be used with ordinary achromatic objectives of large numerical aperture.

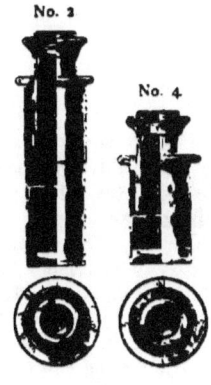

FIG. 32. *Projection Oculars with section removed to show the construction. Below are shown the upper ends with graduated circle to indicate the amount of rotation found necessary to focus the diaphragm on the screen. No. 2, No. 4. The numbers indicate the amount the ocular magnifies the image formed by the objective as with the compensation oculars.* (Zeiss' Catalog, No. 30].

§ 38. **Micrometer Ocular.**—This is an ocular connected with an ocular micrometer. The micrometer may be removable, or it may be permanently in connection with the ocular, and arranged with a spring and screw, by which it may be moved back and forth across the field. (See Ch. IV).

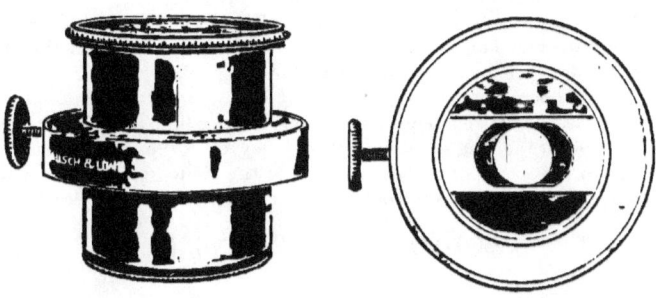

FIG. 33.   FIG. 34.

FIGS. 33-34 *Ocular Micrometer with movable scale. Fig. 33 is a side view of the ocular while Fig. 34 gives a sectional end view, and shows the ocular micrometer in position. In both the screw which moves the micrometer is shown at the left.* (*From Bausch & Lomb Opt. Co.*)

Fig. 35. Ocular Screw-Micrometer with compensation ocular 6. The upper figure shows a sectional view of the ocular and the screw for moving the micrometer at the right. At the left is shown a clamping screw to fasten the ocular to the upper part of the microscope tube. Below is a face view, showing the graduation on the wheel. An ocular micrometer like this is in general like the cob-web micrometer and may be used for measuring objects of varying sizes very accurately. With the ordinary ocular micrometer very small objects frequently fill but a part of an interval of the micrometer, but with this the movable cross lines traverse the object (or rather its real image) regardless of the minuteness of the object. (Zeiss' Catalog, No. 30).

§ 39. **Spectral or Spectroscopic Ocular.**—(See Micro-Spectroscope, Ch. VI).

DESIGNATION OF OCULARS.

§ 40. **Equivalent Focus.**—As with objectives, some opticians designate the oculars by their equivalent focus (§ 13). With this method the power of the ocular, as with objectives, other lenses or lens systems, varies inversely as the equivalent focal length, and therefore the greater the equivalent focal length the less the magnification. This seems as desirable a mode for oculars as for objectives and is coming more and more into use by the most progressive opticians. It is the method of designation advocated by Dr. R. H. Ward for many years, and was recommended by the committee of the American Microscopical Society, (Proc. Amer. Micr. Soc., 1883, p. 175, 1884, p. 228).

§ 41. **Numbering and Lettering.**—Oculars like objectives may be numbered or lettered arbitrarily. When so designated, the smaller the number, or the earlier the letter in the alphabet, the lower the power of the ocular.

§ 42. **Magnification.**—The compensating oculars are marked with the amount they magnify the real image. Thus an ocular marked × 4, indicates that the real image of the objective is multiplied four fold by the ocular.

The projection oculars are designated simply by the amount they multiply the real image of the objective. Thus for the short or 160 mm. tube-length they are, × 2, × 4; and for the long or 250 mm. tube, they are × 3 and × 6. That is, the final image on the screen or the ground glass of the photographic camera will be 2, 3, 4, or 6 times greater than it would be if no ocular were used. See Ch. VIII.

COMPOUND MICROSCOPE.

EXPERIMENTS.

§ 43. **Putting an Objective in Position and Removing it.**—Elevate the tube of the microscope by means of the coarse adjustment (frontispiece) so that there may be plenty of room between its lower end and the stage. Grasp the objective lightly near its lower end with two fingers of the left hand, and hold it against the nut in the lower end of the tube. With two fingers of the right hand take hold of the milled ring near the back or upper end of the objective and screw it into the tube of the microscope. Reverse this operation for removing the objective. By following this method the danger of dropping the objective will be avoided.

§ 44. **Putting an Ocular in Position and Removing it.**—Elevate the body of the microscope with the coarse adjustment so that the objective will be 2 cm. or more from the object—grasp the ocular by the milled ring next the eye-lens (Fig. 21), and the coarse adjustment or the tube of the microscope and gently force the ocular into position. In removing the ocular, reverse the operation. If the above precautions are not taken, and the oculars fit snugly, there is danger in inserting them of forcing the tube of the microscope downward and the objective upon the object.

§ 45. **Putting an Object under the Microscope.**—This is so placing an object under the simple microscope, or on the stage of the compound microscope, that it will be in the field of view when the microscope is in focus (§ 46).

With low powers, it is not difficult to get an object under the microscope. The difficulty increases, however, with the power of the microscope and the smallness of the object. It is usually necessary to move the object in various directions while looking into the microscope, in order to get it into the field. Time is usually saved by getting the object in the center of the field with a low objective before putting the high objective in position. This is greatly facilitated by using a nose-piece, or revolver. (See also § 118, Figs. 61–66).

FIG. 36. *Triple nose-piece or revolver for quickly changing objectives (Queen & Co.).*

§ 46. **Field or Field of View of a Microscope.**—This is the area visible through a microscope when it is in focus. When properly lighted, and there is no object under the microscope, the field appears as a circle of light. When examining an object it appears within the light circle, and by moving the object, if it is of sufficient size, different parts are brought successively into the field of view.

In general, the greater the magnification of the entire microscope, whether the magnification is produced mainly by the objective, the ocular, or by increasing the tube-length, or by a combination of all three (see Ch. IV, under magnification), the smaller is the field.

The size of the field is also dependent, in part, without regard to magnification, upon the size of the opening in the ocular diaphragm. Some oculars, as the orthoscopic and periscopic, are so constructed as to eliminate the ocular diaphragm, and in consequence, although this is not the sole cause, the field is considerably increased. The exact size of the field may be read off directly by putting a stage micrometer under the microscope and noting the number of spaces required to measure the diameter of the light circle.

§ 47. *Table showing the actual size in millimeters of the field of a group of commonly used objectives and oculars. Compare with the graphic representation in Fig. 37. See also § 46.*

Equivalent Focus and N. A. of Objective.	Diameter of Field in mm.		Ocular.
85 mm.	15.4 10.6 8.3	37½ mm. 25 " 12½ "	Huygenian.
45 mm.	7.0 5.0 4.0	37½ mm. 25 " 12½ "	Huygenian.
17 mm.	3.0 2.0 1.6	37½ mm. 25 " 12½ "	Huygenian.
N.A. 0.25	5.7 2.8 1.4 0.97	180 mm. 45 " 15 " 10 "	Compensation.
5 mm.	0.541 0.371 0.290	37½ mm. 25 " 12½ "	Huygenian.
N.A. 0.92	0.850 0.501 0.250 0.173	180 mm. 45 " 15 " 10 "	Compensation.
2 mm.	0.270 0.186 0.147	37½ " 25 " 12½ "	Huygenian.
N.A. = 1.25	0.450 0.251 0.125 0.088	180 mm. 45 " 15 " 10 "	Compensation.

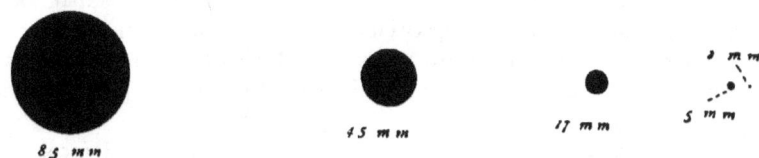

FIG. 37. *Figures showing approximately the actual size of the field with objectives of 85 mm., 45 mm., 17 mm., 5 mm., and 2 mm., equivalent focus, and ocular of 37½ mm., equivalent focus in each case. This figure shows graphically what is also very clearly indicated in the table (§ 47).*

§ 48. The size of the field of the microscope as projected into the field of vision of the normal human eye (*i. e.*, the virtual image) may be determined by the use of the camera lucida with the drawing surface placed at the standard distance of 250 millimeters (Ch. IV).

FUNCTION OF AN OBJECTIVE.

§ 49. Put a 2-in. (50 mm.) objective on the microscope or screw off the front combination of a 16 mm., (⅔-in.), and put the back-combination on the microscope for a low objective.

Place some printed letters or figures under the microscope, and light well. In place of an ocular, put a screen of ground glass, or a piece of lens paper, over the upper end of the tube of the microscope.*

Lower the tube of the microscope by means of the coarse adjustment until the objective is within 2-3 cm. of the object on the stage. Look at the screen on the top of the tube, holding the head about as far from it as for ordinary reading, and slowly elevate the tube by means of the coarse adjustment until the image of the letter appears on the screen.

The image can be more clearly seen if the object is in a strong light and the screen in a moderate light, *i. e.*, if the top of the microscope is shaded.

The letters will appear as if printed on the ground glass or paper, but will be inverted (Fig. 21).

If the objective is not raised sufficiently, and the head is held too near the microscope, the objective will act as a simple microscope. If the letters are erect, and appear to be down in the microscope and not on the screen, hold the head farther from it, shade the screen, and raise the tube of the microscope until the letters do appear on the ground glass.

---

*Ground glass may be very easily prepared by placing some fine emery between two pieces of glass, wetting it with water and then rubbing the glasses together for a few minutes. If the glass becomes too opaque, it may be rendered more translucent by rubbing some oil upon it.

To demonstrate that the object must be outside the principal focus with the compound microscope, remove the screen and turn the tube of the microscope directly toward the sun. Move the tube of the microscope with the coarse adjustment until the burning or focal point is found (§ 6). Measure the distance from the paper object on the stage to the objective, and it will represent approximately the principal focal distance (Figs. 10, 11). Replace the screen over the top of the tube, no image can be seen. Slowly raise the tube of the microscope and the image will finally appear. If the distance between the object and the objective is now taken, it will be found considerably greater than the principal focal distance (compare § 9).

§ 50. **Aerial Image.**—After seeing the real image on the ground-glass, or paper, use the lens paper over about half of the opening of the tube of the microscope. Hold the eye about 250 mm. from the microscope as before and shade the top of the tube by holding the hand between it and the light, or in some other way. The real image can be seen in part as if on the paper and in part in the air. Move the paper so that the image of half a letter will be on the paper and half in the air. Another striking experiment is to have a small hole in the paper placed over the center of the tube opening, then if a printed word extends entirely across the diameter of the tube its central part may be seen in the air, the lateral parts on the paper. The advantage of the paper over part of the opening is to enable one to accommodate the eyes for the right distance. If the paper is absent the eyes adjust themselves for the light circle at the back of the objective, and the aerial image appears low in the tube. Furthermore, it is more difficult to see the aerial image in space than to see the image on the ground-glass or paper, for the eye must be held in the right position to receive the rays projected from the real image, while the granular surface of the glass and the delicate fibers of the paper reflect the rays irregularly, so that the image may be seen at almost any angle, as if the letters were actually printed on the paper or glass.

§ 51. **The function of an objective**, as seen from these experiments, is to form an enlarged, inverted, real image of an object, this image being formed on the opposite side of the objective from the object (Fig. 21).

FUNCTION OF AN OCULAR.

§ 52. Using the same objective as for § 49, get as clear an image of the letters as possible on the lens paper screen. Look at the image with a simple microscope (Fig. 17 or 18) as if the image were an object.

Observe that the image seen through the simple microscope is merely an enlargement of the one on the screen, and that the letters remain inverted, that is they appear as with the naked eye (§ 9). Remove the screen and observe the aerial image with the tripod.

Put a 50 mm., (A, No. 1, or 2 in.) ocular (*i. e.*, an ocular of low magnification) in position (§ 44). Hold the eye about 10 to 20 millimeters from the eye-lens and look into the microscope. The letters will appear as when the simple microscope was used (see above), the image will become more distinct by slightly raising the tube of the microscope with the coarse adjustment.

§ 53. **The Function of the Ocular**, as seen from the above, is that of a simple microscope, viz. : It magnifies the real image formed by the objective as if that image were an object. Compare the image formed by the ocular (Fig. 21), and that formed by a simple microscope (Fig. 38).

It should be borne in mind, however, that the rays from an object as usually examined with a simple microscope, extend from the object in all directions, and no matter at what angle the simple microscope is held, provided it is sufficiently near and points toward the object, an image may be seen. The rays from a real image, however, are continued in

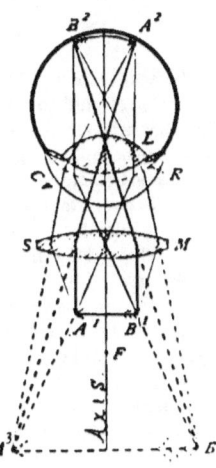

FIG. 38. *Diagram of the simple microscope showing the course of the rays and all the images, and that the eye forms an integral part of it.*

$A^1 B^1$. *The object within the principal focus.* $A^1 B^3$. *The virtual image on the same side of the lens as the object. It is indicated with dotted lines, as it has no actual existence.*

$B^2 A^2$. *Retinal image of the object ($A^1 B^1$). The virtual image is simply a projection of the retinal image in the field of vision.*

Axis. The principal optic axis of the microscope and of the eye. Cr. Cornea of the eye. L. Crystalline lens of the eye. R. Ideal refracting surface at which all the refractions of the eye may be assumed to take place.

certain definite lines and not in all directions ; hence, in order to see this aerial image with an ocular or simple microscope, or in order to see the aerial image with the unaided eye, the simple microscope, ocular or eye must be put in the path of the rays (Fig. 21).

§ 54. **The field-lens** of a Huygenian ocular makes the real image smaller and consequently increases the size of the field ; it also makes the image brighter by contracting the area of the real image (Fig. 30).

Demonstrate this by screwing off the field-lens and using the eye-lens alone as in the ocular, refocusing if necessary. Note also that the letters or other image is bordered by a colored haze (§ 7).

When looking into the ocular with the field-lens removed, the eye should not be held so close to the ocular, as the eye-point is considerably farther away than when the field-lens is in place.

§ 55. **The eye-point.**—This is the point above the ocular or simple microscope where the greatest number of emerging rays cross. Seen in profile, it may be likened to the narrowest part of an hour glass. Seen in section (Fig. 30), it is the smallest and brightest light circle above the ocular. This is called the eye-point, for if the pupil of the eye is placed at this level, it will receive the greatest number of rays from the microscope, and consequently see the largest field.

Demonstrate the eye-point by having in position an objective and ocular as above (§ 49). Light the object brightly, focus the microscope, shade the ocular, then hold some ground-glass or a piece of the lens paper above the ocular and slowly raise and lower it until the smallest circle of light is found. By using different oculars it will be seen that the eye-point is nearer the eye-lens in high than in low oculars, that is the eye-point is nearer the eye-lens for an ocular of small equivalent focus than for one of greater focal length.

REFERENCES FOR CHAPTER I.

In the appendix will be given a bibliography, with full titles, of the works and periodicals referred to.

For the subjects considered in this chapter general works on the microscope may be consulted with great advantage for different or more exhaustive treatment. The most satisfactory work in English is Carpenter-Dallinger. For the history of the microscope, Mayall's Cantor Lectures on the microscope are very satisfactory. See also Beale, E Bausch, Beherens, Kossel and Schiefferdecker, Dippel, Frey. Harting, Hogg, Nägeli and Schwendener, Robin, Van Heurck, Clark, Cross and Cole, Stokes.

The following special articles in periodicals may be examined with advantage:

Apochromatic Objectives, etc. Dippel in Zeit. wiss. Mikr., 1886, p. 303; also in the Jour. Roy. Micr. Soc., 1886, pp. 316, 849, 1110; same, 1890, p. 480; Zeit. f. Instrumentenk., 1890, pp. 1-6; Micr. Bullt., 1891, pp. 6-7.

Tube-length, etc. Gage, Proc. Amer. Soc. Micrs., 1887, pp. 168-172; also in the Microscope, the Jour. Roy. Micr. Soc., and in Zeit. wiss. Mikr., 1887-8, Bausch, Proc. Amer. Soc. Micrs., 1890, pp. 43-49; also in the Microscope, 1890, pp. 289-296.

Aperture. J. D. Cox, Presidential Address, Proc. Amer. Soc. Micrs., 1884, pp. 5-39, Jour. Roy. Micr. Soc., 1881, pp. 303, 348, 365, 388; 1882, pp. 300, 460; 1883, p. 790; 1884, p. 20.

CHAPTER II.

LIGHTING AND FOCUSING; MANIPULATION OF DRY, ADJUSTABLE AND IMMERSION OBJECTIVES; CARE OF THE MICROSCOPE AND OF THE EYES; LABORATORY MICROSCOPES.

APPARATUS AND MATERIAL FOR THIS CHAPTER.

Microscope supplied with plane and concave mirror, achromatic and Abbe condensers, dry, adjustable and immersion objectives, oculars, tripple nose-piece. Microscope lamp and movable condenser (bull's eye or other form (Fig. 52), Homogeneous immersion liquid; Benzin, alcohol, distilled water; Mounted preparation of fly's wing (§ 68); Mounted preparation of *Pleurasigma*. Stage or ocular micrometer with lines filled with graphite (§ 73, 74); Glass slides and cover-glasses (Ch. VII); 10 per ct. solution of salicylic acid in 95 per ct. alcohol (§ 88); Preparation of stained microbes (§ 101); Vial of equal parts olive or cotton seed oil or liquid vaselin and benzin (§ 105); Ward's and double eye-shade (Figs. 59, 60); Screen for whole microscope (Fig. 58).

FOCUSING.

§ 56. **Focusing** is mutually arranging an object and the microscope so that a clear image may be seen.

With a simple microscope (§ 9) either the object or the microscope or both may be moved in order to see the image clearly, but with the compound microscope the object more conveniently remains stationary on the stage, and the tube or body of the microscope is raised or lowered (frontispiece).

In general, the higher the power of the whole microscope whether simple or compound, the nearer together must the object and objective be brought. With the compound microscope, the higher the objective, and the longer the tube of the microscope, the nearer together must the object and the objective be brought. If the oculars are not par-focal, the higher the magnification of the ocular, the nearer must object and objective be brought.

§ 57. **Working Distance.**— By this is meant the space between the simple microscope and the object, or between the front lens of the compound microscope and the object, when the microscope is in focus. This working distance is always considerably less than the equivalent focal length of the objective. For example, the front-lens of a ¼th in., or 6 mm. objective would not be ¼th inch, or 6 millimeters from the object when the microscope is in focus, but considerably less than that distance. If there were no other reason than the limited working distance of high objectives, it would be necessary to use very thin cover-glasses over the object. (See ¿ 22, 27). If too thick covers are used, it may be impossible to get an objective near enough an object to get it in focus. For objects that admit of examination with high powers it is always better to use thin covers.

LIGHTING WITH DAYLIGHT.

§ 58. Unmodified sunlight should not be employed except in special cases. North light is best and most uniform. When the sky is covered with white clouds the light is most favorable. To avoid the shadows produced by the hands in manipulating the mirror, etc., it is better to face the light ; but to protect the eyes and to shade the stage of the microscope some kind of screen should be used. The one figured in (Fig. 58) is cheap and efficient. If one dislikes to face the window or lamp it is better to sit so that the light will come from the left as in reading.

It is of the greatest importance and advantage for one who is to use the microscope for serious work that he should comprehend and appreciate thoroughly the various methods of illumination, and the special appearances due to different kinds of illumination.

Depending on whether the light illuminating an object traverses the object or is reflected upon it, and also whether the object is symmetrically lighted, or lighted more on one side than the other, light used in microscopy is designated as *reflected and transmitted, axial and oblique.*

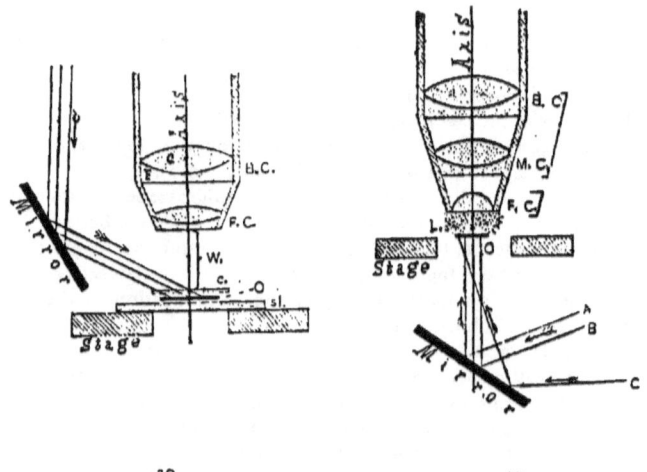

FIGS. 39–40. *For full explanation see Figs. 22 and 23.*

§ 59. **Reflected, Incident or Direct Light.**—By this is meant light reflected upon the object in some way and then irregularly reflected from the object to the microscope. By this kind of light objects are ordinarily seen by the unaided eye, and the objects are mostly opaque. In Vertebrate Histology, reflected light is but little used; but in the study of opaque objects, like whole insects, etc., it is used a great deal. For low powers, ordinary daylight that naturally falls upon the object, or is reflected or condensed upon it with a mirror or condensing lens, answers very well. For high powers and for special purposes, special illuminating apparatus has been devised (§ 26). (See also Carpenter-Dallinger, p. 278).

§ 60. **Transmitted Light.**—By this is meant light which passes through an object from the opposite side. The details of a photographic negative are in many cases only seen or best seen by transmitted light, while the print made from it is best seen by reflected light.

Almost all objects studied in Vertebrate Histology are lighted by transmitted light, and they are in some way rendered transparent or semi transparent. The light traversing and serving to illuminate the object in working with a compound microscope is usually reflected from a plane or concave mirror, or from a mirror to a condenser (§ 75), and thence transmitted to the object from below (Figs. 48-51).

§ 61. **Axial or Central Light.**—By this is understood light reaching the object, the rays of light being parallel to each other and to the optic axis of the microscope, or a diverging or converging cone of light whose axial ray is parallel with the optic axis of the microscope. In either case the object is symmetrically illuminated.

§ 62. **Oblique Light.**—This is light in which parallel rays from a plane mirror form an angle with the optic axis of the microscope (Fig. 40). Or if a concave mirror or a condenser is used, the light is oblique when the axial ray of the cone of light forms an angle with the optic axis (Fig. 49).

DIAPHRAGMS.

§ 63. **Diaphragms and their Proper Employment.**—Diaphragms are opaque disks with openings of various sizes, which are placed between the source of light or mirror and the object. In some cases an iris diaphragm is used, and then the same one is capable of giving a large range of openings. The object of a diaphragm, in general, is to cut off all adventitious light and thus enable one to light the object in such a way that the light finally reaching the microscope shall all come from the object or its immediate vicinity. The diaphragms of a condenser serve to vary its aperture to the needs of each object and each objective.

§ 64. **Size and Position of Diaphragm Opening.**—When no condenser is used the size of the opening in the diaphragm should be about that of the front lens of the objective used. For some objects and some objectives this rule may be quite widely departed from; one must learn by trial.

When lighting with a mirror the diaphragm should be as close as possible to the object in order, (a) that it may exclude all adventitious light from the object; (b) that it may not interfere with the most efficient illumination from the mirror by cutting off a part of the illuminating pencil. If the diaphragm is a considerable distance below the object, (1) it allows considerable adventitious light to reach the object and thus injures the distinctness of the microscopic image; 2) it prevents the use of very oblique light unless it swings with the mirror; (3) it cuts off a part of the illuminating cone from a concave mirror. On the other hand, even with a small diaphragm, the whole field will be lighted.

With an illuminator or condenser (Figs. 41, 48), the diaphragm serves to narrow the pencil to be transmitted through the condenser, and thus to limit the aperture or for any special purpose to be served (see § 80). Furthermore, by making the diaphragm opening eccentric, oblique light may be used, or by using a diaphragm with a slit around the edge (central stop diaphragm), the center remaining opaque, the object may be lighted with a hollow cone of light, all of the rays having great obliquity. In this way the so called dark-ground illumination may be produced (§ 88; Fig. 51).

ARTIFICIAL ILLUMINATION.

§ 65. For evening work and for certain special purposes, artificial illumination is employed. A good petroleum (kerosene) lamp with flat wick has been found very satisfactory, but for brilliancy and for the actinic power necessary for photomicrography (see Ch. VIII) the new acetylene light seems to be all that could be desired. Whatever source of artificial light is employed, the light should be brilliant and steady.

LIGHTING AND FOCUSING : EXPERIMENTS.

§ 66. **Lighting with a Mirror.**—Place a mounted fly's wing under the microscope, put the 16 mm. (⅔ in.) or other low objective in position, also a low ocular. With the coarse adjustment, lower the tube of the microscope to within about 1 cm. of the object. Use an opening in the diaphragm about as large as the front lens of the objective; then with the plane mirror try to reflect light up through the diaphragm upon the object. One can tell when the field (§ 46) is illuminated, by looking at the object on the stage, but more satisfactorily by looking into the microscope. It sometimes requires considerable manipulation to light the field well. After using the plane side of the mirror turn the concave side into position and light the field with it. As the concave mirror condenses the light, the field will look brighter with it than with the plane mirror. It is especially desirable to remember that the excellence of lighting depends in part on the position of the diaphragm (§ 57). If the greatest illumination is to be obtained from the concave mirror, its position must be such that its focus will be at the level of the object. This distance can be very easily determined by finding the focal point of the mirror in full sunlight.

§ 67. **Use of the Plane and of the Concave Mirror.**—The mirror should be freely movable, and have a plane and a concave face. The concave face is used when a large amount of light is needed, the plane face when a moderate amount is needed or when it is necessary to have parallel rays or to know the direction of the rays.

§ 68. **Focusing with Low Objectives.**— Place a mounted fly's wing under the microscope; put the 16 mm. (⅔ in.) objective in position, and also the lowest ocular. Select the proper opening in the diaphragm and light the object well with transmitted light (§ 60, 64).

Hold the head at about the level of the stage, look toward the window, and between the object and the front of the objective; with the coarse adjustment lower the tube until the objective is within about half a cm. of the object. Then look into the microscope and slowly elevate the tube with the coarse adjustment. The image will appear dimly at first, but will become very distinct by turning the tube still higher. If the tube is raised too high the image will become indistinct, and finally disappear. It will again appear if the tube is lowered the proper distance.

When the microscope is well focused try both the concave and the plane mirrors, in various positions and note the effect. Put a high ocular in place of the low one (§ 40). If the oculars are not par-focal it will be necessary to lower the tube somewhat to get the image in focus.*

Pull out the draw-tube 4–6 cm., thus lengthening the body of the microscope, and it will be found necessary to lower the tube of the microscope somewhat. (For reason, see Fig. 57).

§ 69. **Pushing in the Draw-Tube.** — To push in the draw-tube, grasp the large milled ring of the ocular with one hand, and the milled head of the coarse adjustment with the other, and gradually push the draw-tube into the tube. If this were done without these precautions the objective might be forced against the object and the ocular thrown out by the compressed air.

§ 70. **Focusing with High Objectives.** — Employ the same object as before, elevate the tube of the microscope and remove the 16 mm. (⅔ in.) objective as indicated. Put the 3 mm. (⅛ in.) or a higher objective in place, and use a low ocular.

Light well, and employ the proper opening in the diaphragm, etc. (§ 64). Look between the front of the objective and the object as before (§ 68), and lower the tube with the coarse adjustment till the objective almost touches the cover-glass over the object. Look into the microscope, and with the coarse adjustment, raise the tube very slowly until the image begins to appear, then turn the milled head of the fine adjustment (frontispiece), first one way and then the other, if necessary, until the image is sharply defined.

In practice it is found of great advantage to move the preparation slightly while focusing. This enables one to determine the approach

---

* Par-focal oculars are so constructed, or so mounted, that those of different powers may be interchanged without the microscopic image becoming wholly out of focus (Fig. 31, note, p. 23). When high objectives are used, while the image may be seen after changing oculars, the instrument nearly always needs slight focusing. With low powers this may not be necessary.

to the focal point either from the shadow or the color, if the object is colored. With high powers and scattered objects there might be no object in the small field (see § 46, Fig. 37, for size of field). By moving the preparation an object will be moved across the field and its shadow gives one the hint that the objective is approaching the focal point. It is sometimes desirable to focus on the edge of the cement ring or on the little ring made by the marker (see Figs. 61–65 § 118).

Note that this high objective must be brought nearer the object than the low one, and that by changing to a higher ocular (if the oculars are not par-focal) or lengthening the tube of the microscope it will be found necesssary to bring the objective still nearer the object, as with the low objective. (For reason see Fig. 57).

§ 71. **Always Focus Up,** as directed above. If one lowers the tube only when looking at the end of the objective as directed above, there will be no danger of bringing the objective in contact with the object, as may be done if one looks into the microscope and focuses down.

When the instrument is well focused, move the object around in order to bring different parts into the field of view (§ 46). It may be necessary to re-focus with the fine adjustment every time a different part is brought into the field. In practical work, one hand is kept on the fine adjustment constantly, and the focus is continually varied.

§ 72. **Determination of Working Distance.**—As stated in § 57 this is the distance between the front lens of the objective and the object when the objective is in focus. It is always less than the equivalent focal length of the objective.

Make a wooden wedge 10 cm. long which shall be exceedingly thin at one end and about 20 mm. thick at the other. Place a slide on the stage and some dust on the slide. Do not use a cover-glass. Focus the dust carefully first with the low then with the high objective. When the objective is in focus push the wedge under the objective on the slide until it touches the objective. Mark the place of contact with a pencil and then measure the thickness of the wedge with a rule opposite the point of contact. This thickness will represent very closely the working distance. For measuring the thickness of the wedge at the point of contact for the high objective use a steel scale ruled in $\frac{1}{4}$ths mm. and the tripod to see the divisions. Or one may use a cover-glass measurer, for determining the thickness of the wedge (Ch. VIII).

For the higher powers, if one has a microscope in which the fine adjustment is graduated, the working distance may be readily determined when the thickness of the cover-glass over the specimen is known, as follows: Get the object in focus, lower the tube of the microscope until the front of the objective just touches the cover-glass. Note the position of the micrometer screw and slowly focus up with the fine adjustment until the object is in focus. The distance the objective was raised plus the thickness of the cover-glass represents the working distance. For example, a 3 mm. objective after being brought in contact with a cover-glass was raised by the fine adjustment a distance represented by 16 of the divisions on the head of the micrometer screw. Each division represented .01 mm., consequently the objective was raised .16 mm. As the cover-glass on the specimen used was .15 mm. the total working distance is .16 + .15 = .31 mm.

CENTRAL AND OBLIQUE LIGHT WITH A MIRROR.

§ 73. **Axial or Central Light** (§ 61).—Remove the condenser or any diaphragm from the substage, then place a preparation containing minute air bubbles under the microscope. The preparation may be easily made by beating a drop of mucilage on a slide and covering it. (See Ch. III). Use a 3 mm. (⅛ in.) or No. 7 objective and a medium ocular. Focus the microscope and select a very small bubble, one whose image appears about 1 mm. in diameter, then arrange the plane mirror so that the light spot in the bubble appears exactly in the center. Without changing the position of the mirror in the least, replace the air-bubble preparation by one of *Pleurasigma angulatum* or some other finely marked diatom. Study the appearance very carefully.

§ 74. **Oblique Light,** (§ 62).—Swing the mirror far to one side so that the rays reaching the object may be very oblique to the optic axis of the microscope. Study carefully the appearance of the diatom with the oblique light. Compare the different appearance with that of central light. The effect of oblique light is not so striking with histological preparations as with diatoms.

It should be especially noted in §§ 73, 74, that one cannot determine the exact direction of the rays by the position of the mirror. This is especially true for axial light (§ 73). To be certain that the light is axial some such test as that given in § 73 should be applied. (See also Ch. III, under Air-bubbles).

CONDENSERS OR ILLUMINATORS.*

§ 75. These are lenses or lens-systems for the purpose of illuminating with transmitted light the object to be studied with the microscope.

---

*No one has stated more clearly or appreciated more truly the value of correct illumination and the methods of obtaining it than Sir David Brewster, 1820, 1831.

For the highest kind of investigation their value cannot be overestimated. They may be used either with natural or artificial light, and should be of sufficient numerical aperture to satisfy objectives of the widest angle.

It is of the greatest advantage to have the sub-stage condenser mounted with rack and pinion so that it may be easily moved up or down under the stage. The iris diaphragm is so convenient that it should be furnished in all cases, and there should be marks indicating the N. A. of the condenser utilized with different openings. Finally, the condenser should be supplied with central stops for dark-ground illumination (§ 88) and with blue and neutral tint glasses to soften the glare when artificial light is used (§§ 85, 89).

Condensers or **Illuminators** fall into two great groups, the **Achromatic**, giving a large aplanatic cone, and **Non-achromatic**, giving much light, but a relatively small aplanatic cone of light.

§ 76. **Achromatic Condenser.**—It is still believed by all expert microscopists that the contention of Brewster was right, and the condenser to give the greatest aid in elucidating microscopic structure must approach in excellence the best objectives. That is, it should be as free as possible from spherical and chromatic aberration, and therefore would transmit to the object a very large aplanatic cone of light. Such condensers are especially recommended for photo-micrography by all, and those who believe in getting the best possible image in every case are equally strenuous that achromatic condensers should be used for all work. Unfortunately good condensers like good objectives are expensive, and student microscopes as well as many others are mostly supplied with the non-achromatic condensers or with none.

Many excellent achromatic condensers have been made, but the most perfect of all seems to be the apochromatic of Powell and Lealand (Car-

---

He says of illumination in general : "The art of illuminating microscopic objects is not of less importance than that of preparing them for observation." "The eye should be protected from all extraneous light, and should not receive any of the light which proceeds from the illuminating center, excepting that portion of it which is transmitted through or reflected from the object." So likewise the value and character of the substage condenser was thoroughly understood and pointed out by him as follows : "I have no hesitation in saying that the apparatus for illumination requires to be as perfect as the apparatus for vision, and on this account I would recommend that the illuminating lens should be perfectly free of chromatic and spherical aberration, and the greatest care be taken to exclude all extraneous light both from the object and from the eye of the observer." See Sir David Brewster's Treatise on the Microscope, 1837, pp. 136, 138, 146, and the Edinburgh *Journal of Science*, new series, No. 11 (1831), p. 83.

penter-Dallinger, pp. 254, 263). To attain the best that was possible many workers have adopted the plan of using objectives as condensers. A special substage fitting is provided with the proper screw and the objective is put into position, the front lens being next the object. As will be seen below (§ 79-80), the full aperture of an objective can rarely be used, and for histological preparations perhaps never, so that an objective of greater equivalent focus, *i. e.*, lower power is used than the one on the microscope. It is much more convenient, however, to have a special condenser with iris diaphragm or special diaphragms so that one may use any aperture at will, and thus satisfy the conditions necessary for lighting different objects for the same objective and for lighting with objectives of different apertures. An excellent condenser of this form has been produced by Zeiss (Fig. 41). It has a total numerical aperture of 1.00, and an aplanatic aperture of N. A. 0.65.

FIG. 41. *Zeiss' Achromatic Condenser. c.s c s. Centering screws for changing the position of the condenser and making its axis continuous with that of the microscope. A segment of the condenser is cut away to show the combinations of lenses. For very low powers the upper lens is sometimes screwed off. There is an iris diaphragm between the middle and lower combinations. (Zeiss' Catalog, No. 30)*

§ 77. **Centering the Condenser.**—To get the best possible illumination for bringing out in the clearest manner the minute details of a microscopic object two conditions are necessary, viz. : The principal optic axis of the condenser must be continuous with that of the microscope (see frontispiece) and the object must be in the focus of the condenser, *i. e.*, at the apex of the cone of light given by the condenser.

The centering is most conveniently accomplished as follows although daylight may be used with almost equal facility : The object is placed on the stage and lighted with the edge or face of the flame and then a very small diaphragm is put below the condenser. (If the Zeiss achromatic condenser is used, the diaphragm of the Abbe illuminator serves for this. If there is no pin-hole diaphragm one can be made of stiff, black paper, care must be taken, however, to make the opening exactly central. This is best accomplished by putting the paper disc over the iris or metal diaphragm and then making the hole in the center of the small circle uncovered by the metal diaphragm. For the hole a fine needle is best). If now the condenser is lowered or racked away from the objective the image of the diaphragm will appear. If the opening is not central it should be made so by using the centering screws of the condenser.

A better plan than to lower the condenser to focus the image of the diaphragm, is to raise the body of the microscope slowly with the coarse adjustment. It is almost impossible to make apparatus so accurate that two parts like the body of the microscope and the substage, each working on different sliding surfaces, shall continue in exactly the same plane. So one will find that if the condenser be accurately centered with the condenser lowered, and then the condenser be racked up close to the stage and the image of the diaphragm opening brought again into focus by racking up the body of the microscope, it will not be found accurately centered in most cases. For this reason it is advised that the condenser be left in position close to the stage and the tube of the microscope be used to focus the diaphragm exactly as in ordinary work.

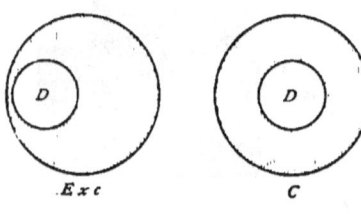

FIG. 42. Shows that the optic axis of the condenser does not coincide with that of the microscope. (D). Diaphragm of the condenser shown at one side of the field of the microscope.

FIG. 43 Shows the diaphragm (D) in the center of the field of the microscope, and thus the coincidence of the axis of the condenser with that of the microscope.

FIG. 42.   FIG. 43.

§ 78. **Centering the Image of the Source of Illumination.**—For the best results it is not only necessary that the condenser be properly centered, but that the object to be studied should be in the image of the source of illumination and that this should also be centered (Figs. 44, 45). After the condenser itself is centered the iris diaphragm is opened to its full extent or the diaphragm carrier turned wholly aside. The condenser is then racked up toward the objective until the image of the flame is apparently on the specimen. If this cannot be accomplished the relative position of the lamp and condenser is not correct and should be so changed that the image of the edge of the flame is sharply defined. This image must also be centered. This is easily accomplished by manipulation of the mirror or, if a lamp is used, by changing the position of the lamp or of the bull's eye (Figs. 34, 52).

CH. II.] LIGHTING AND FOCUSING. 43

§ 79. Proper Numerical Aperture of the Condenser.—As stated above, the aperture of the condenser should have a range by means of properly selected diaphragms to meet the requirements of all objectives from the lowest to those of the highest aperture. It is found in practice that for diatoms, etc., the best images are obtained when the object is lighted with a cone which shall fill about three-fourths of the diameter of the back lens of the objective with light, but for histological and other preparations of lower refractive power only one-half or one-third the aperture can be utilized.

Fig. 44 Shows the image of the flame (Fl.) in the center (C) of the field of the microscope and illuminating the object.

Fig. 45. Shows the image of the flame (Fl.) at one side of the center (Exc.) and not properly illuminating the object.

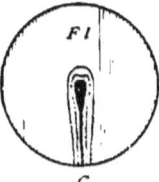
FIG. 44.

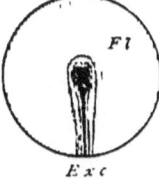
FIG. 45.

To determine this in any case, focus the object carefully, take out the ocular, look down the tube at the back lens. If less than three-fourths of the back lens is lighted, increase the opening in the diaphragm—if more than three-fourths, diminish it. For some objects it is advantageous to use the entire aperture, for others, less than three-fourths. Experience will teach the best lighting for special cases.

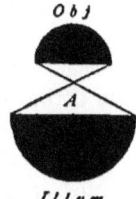
FIG. 46.

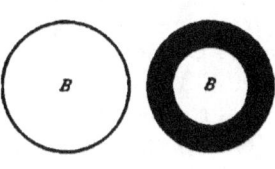
FIG. 47.

FIGS. 46-47. Figures showing the dependence of the objective upon the illuminating cone of the condenser. (Nelson.)

FIG. 46 (A). The illuminating cone from the condenser (Illum). This is seen to be just sufficient to fill the objective (Obj.).

(B). The back lens of the objective entirely filled with light, showing that the numerical aperture of the illuminator is equal to that of the objective.

FIG. 47. (A). In this figure the illuminating cone from the condenser (Illum.) is seen to be insufficient to fill the objective (Obj.).

(B). The back lens of the objective only partly filled with light, due to the restricted aperture of the illuminator.

§ 80. **Aperture of the Illuminating Cone and the Field.**—It is to be remarked that with a very small source of light the entire aperture of the objective may be filled if a proper illuminator or condenser is used. The aperture depends on the diaphragm used with the condenser. And the size of the diaphragm must be directly as the aperture of the objective. That is, it is just the reverse of the rule for diaphragms where no condenser is used (§ 63); for there the diaphragm is made large for low powers, and consequently low apertures, while with the condenser the diaphragm is made small for low and large for high powers as the aperture is greater in the high powers of a given series of objectives. It is very instructive to demonstrate this by using a 16 mm. objective and opening the diaphragm of the condenser till the back lens is just filled with light. Then if one uses a 3 or 4 mm. objective it will be seen that the back lens of the higher objective is only partly filled with light, and to fill it the diaphragm must be much more widely opened.

With a condenser, then, the diaphragm has simply to regulate the aperture of the illuminating cone, and has nothing to do with lighting a large or a small field.

With the condenser, there are two conditions that must be fulfilled,— the proper aperture must be used, and that is determined by the diaphragm, and secondly the whole field must be lighted. The latter is accomplished by using a larger source of light, as the face instead of the edge of a lamp flame, or by lowering or raising the condenser so that the object is not in the focus of the condenser, but above or below it, and therefore lighted by a converging or diverging beam where the light is spread over a greater area (Figs. 48–51).

§ 81. **Non-Achromatic Condenser.**—Of the non-achromatic condensers or illuminators, the Abbe condenser or illuminator is the one most generally used. It is also much more commonly used than the achromatic condenser from its cheapness. It consists of two or three very large lenses and transmits a cone of light of 1.20 N. A. to 1.40 N. A., but the aberrations, both spherical and chromatic, are very great in both forms. Indeed, so great are they that in the best form of three lenses with an illuminating cone of 1.40 N. A., the aplanatic cone transmitted is only 0.5, and it is the aplanatic cone which is of real use in microscopic illumination where details are to be studied. There is no doubt, however, that the results obtained with a non-achromatic condenser like the Abbe are much more satisfactory than with no condenser. The highest results cannot be attained with it, however. (Carpenter-Dallinger, p. 256).

§ 82. **Arrangement of the Condenser.**—The proper position of the illuminator for high objectives is one in which the beam of light traversing it is brought to a focus on the object. If parallel rays are reflected from the plane mirror to it, they will be focused only a few millimeters above the upper lens of the condenser; consequently the illuminator should be about on the level of the top of the stage and therefore almost in contact with the lower surface of the slide. For some purposes, when it is desirable to avoid the loss of light by reflection or refraction, a drop of water or homogeneous immersion fluid is put between the slide and condenser, forming the so-called immersion illuminator. This is necessary only with objectives of high power and large aperture or for dark-ground illumination.

§ 83. **Centering the Condenser.**—The illuminator should be centered to the optic axis of the microscope, that is the optic axis of the condenser and of the microscope should coincide. Unfortunately there is extreme difficulty in determining when the Abbe illuminator is centered. Centering is approximated as follows: Put a pin-hole diaphragm over the end of the condenser (Fig. 48)—that is, a diaphragm with a small central hole—the central opening should appear to be in the middle of the field of the microscope. If it does not, the condenser should be moved from side to side by loosening the centering screws until it is in the center of the field. In case no pin-hole diaphragm accompanies the condenser, one may put a very small drop of ink, as from a pen-point, on the center of the upper lens and look at it with the microscope to see if it is in the center of the field. If it is not, the condenser should be adjusted until it is. When the condenser is centered as nearly as possible remove the pin-hole diaphragm or the spot of ink. The microscope and illuminator axes may not be entirely coincident even when the center of the upper lens appears in the center of the field, as there may be some lateral tilting of the condenser, but the above is the best the ordinary worker can do, and unless the mechanical arrangements of the illuminator are very deficient, it will be very nearly centered.

It is to be hoped that the opticians will devise some kind of mounting for this the most commonly used condenser whereby it may be centered as described for the achromatic condenser instead of by the crude methods described above. If the condenser mounting regularly possessed centering screws as in the microscope of Watson & Sons (Fig. 71), and there was a centering diaphragm in the proper position so that its image could be projected into the field of view, the operation would be very simple. If, further, the condensers of Powell and Lealand were selected as models the condensers need not be so bulky, and still retain all their efficiency.

§ 84. **Mirror and Light for the Abbe Condenser.**—It is best to use light with parallel rays. The rays of daylight are practically parallel; it is best, therefore, to employ the plane mirror for all but the lowest powers. If low powers are used the whole field might not be illuminated with the plane mirror when the condenser is close to the object; furthermore, the image of the window frame, objects outside the building, as trees, etc., would appear with unpleasant distinctness in the field of the microscope. To overcome these defects, one can lower the condenser and thus light the object with a diverging cone of light, or use the concave mirror and attain the same end when the condenser is close to the object (Fig. 48).

§ 85. **Artificial Light.**—If one uses lamplight, it is recommended that a large bull's eye be placed in such a position between the light and the mirror that parallel rays fall upon the mirror or in some cases an image of the lamp flame. If one does not have a bull's eye the concave mirror may be used to render the rays less divergent. It may be necessary to lower the illuminator somewhat in order to illuminate the object in its focus.

ABBE CONDENSER : EXPERIMENTS.

§ 86. **Abbe Condenser, Axial and Oblique Light.**—Use a diaphragm a little larger than the front lens of the 3 mm. (⅛ in.) objective, have the illuminator on the level, or nearly on the level, of the upper surface of the stage, and use the plane mirror. Be sure that the diaphragm carrier is in the notch indicating that it is central in position. Use the *Pleurasigma* as object. Study carefully the appearance of the diatom with this central light, then make the diaphragm eccentric so as to light with oblique light. The differences in appearance will probably be even more striking than with the mirror alone (§ 74).

§ 87. **Lateral Swaying of the Image.**—Frequently in studying an object, especially with a high power, it will appear to sway from side to side in focusing up or down. A glass stage micrometer or fly's wing is an excellent object. Make the light central or axial and focus up and down and notice that the lines simply disappear or grow dim. Now make the light oblique, either by making the diaphragm opening eccentric or if simply a mirror is used, by swinging the mirror sidewise. On focusing up and down, the lines will sway from side to side. What is the direction of apparent movement in focusing down with reference to the illuminating ray ? What in focusing up ? If one understands this experiment it may sometimes save a great deal of confusion. (See under testing the microscope for swaying with central light § 104).

§ 88. **Dark-Ground Illumination.**—When an object is lighted with rays of a greater obliquity than can get into the front lens of the objective, the field will appear dark (Fig. 51). If now the object is com-

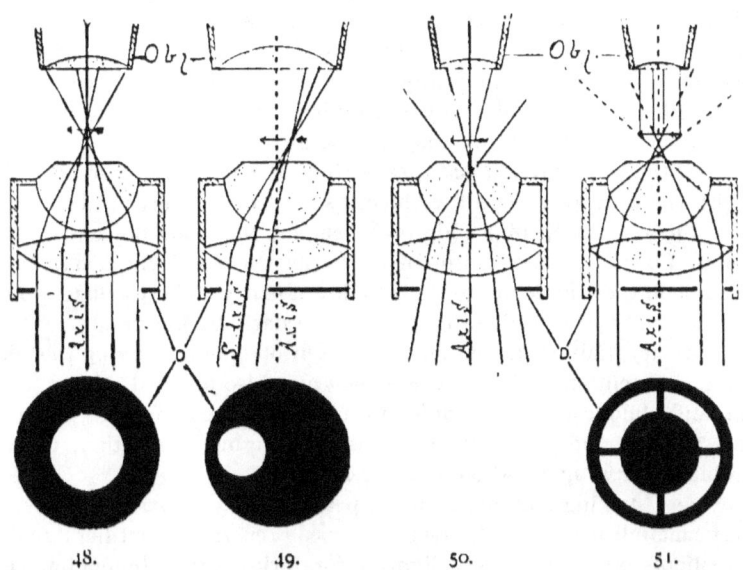

FIGS. 48-51. *Sectional views of the Abbe Illuminator of 1.20 N. A. showing various methods of illumination (§ 84).* Fig. 48, *axial light with parallel rays.* Fig. 49, *oblique light.* Fig. 50, *axial light with converging beam.* Fig. 51, *dark-ground illumination with a central stop diaphragm.*
Axis. The optic axis of the illuminator and of the microscope. The illuminator is centered, that is its optic axis is a prolongation of the optic axis of the microscope.
S. Axis. Secondary axis. In oblique light the central ray passes along a secondary axis of the illuminator, and is therefore oblique to the principal axis.
D D. Diaphragms. These are placed in sectional and in face views. The diaphragm is placed between the mirror and the illuminator. In Fig. 49 the opening is eccentric for oblique light, and in Fig. 51 the opening is a narrow ring, the central part being stopped out, and thus giving rise to dark-ground illumination (§ 88).
Obj. Obj. The front of the objective.

posed of fine particles, or is semi-transparent, it will refract or reflect the light which meets it, in such a way that a part of the very oblique rays will pass into the objective, hence as light reaches the objective only from the object, all the surrounding field will be dark and the object will appear like a self-luminous one on a dark back ground. This form of illumination is only successful with low powers and objectives of small aperture. It is well to make the illuminator immersion for this experiment, see § 98.

(A) *With the Mirror*—Remove all the diaphragms so that very oblique light may be used, employ a stage micrometer in which the lines have been filled with graphite, use a 16 mm. (⅔ in.) objective, and when the light is sufficiently oblique the lines will appear something like streaks of silver on a black back-ground. A specimen like that described below in (B) may also be used.

(B) *With the Abbe Condenser.*—Have the illuminator so that the light would be focused on the object (see § 82) and use a diaphragm with the annular opening (Fig. 51); employ the same objective as in (A). For object place a drop of 10 % solution of salicylic acid in 95 % alcohol on the middle of a slide and allow it to dry and crystallize. The crystals will appear brilliantly lighted on a dark back-ground. Put in an ordinary diaphragm and make the light oblique by making the diaphragm eccentric. The same specimen may also be tried with a mirror and oblique light. In order to appreciate the difference between this dark-ground and ordinary transmitted-light illumination, use an ordinary diaphragm and observe the crystals.

A very striking and instructive experiment may be made by adding a very small drop of the solution to the dried preparation, putting it under the microscope very quickly, lighting for dark-ground illumination and then watching the crystallization.

ARTIFICIAL ILLUMINATION.

§ 89. For evening work and for regions where daylight is not sufficiently brilliant, artificial illumination must be employed. Furthermore, for the most critical investigation of bodies with fine markings like diatoms, artificial light has been found superior to daylight.

A petroleum (kerosene) lamp with flat wick gives a satisfactory light. It is recommended that instead of the ordinary glass chimney one made of metal with a slit-like opening covered with an oblong cover-glass is more satisfactory, as the source of light is more restricted. Very excellent results may be obtained, however, with the ordinary bed-room lamp furnished with the usual glass chimney.

The new acetylene light promises to be the most perfect of all the artificial lights for microscopic observation and for photo-micrography. (See under Photo-micrography).

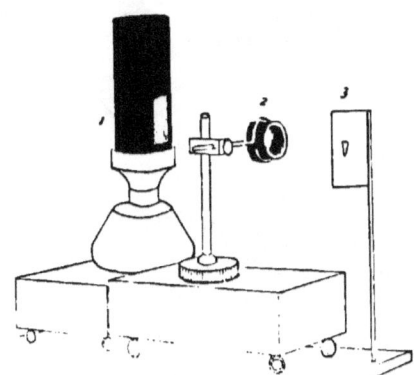

FIG. 52. 1. *Lamp with slit-opening in metal chimney.* 2. *Bull's eye on separate stand.* 3. *Screen showing image of flame.*

Whenever possible, the edge of the flame is turned toward the microscope, the advantage of this arrangement is the greater brilliancy, due to the greater thickness of the flame in this direction.

§ 90. **Mutual Arrangement of Lamp, Bull's Eye and Microscope.**—To fulfill the conditions given above, namely, that the object be illuminated by the image of the source of illumination the lamp must be in such a position that the condenser projects a sharp image of the flame upon the object (Fig. 52), and only by trial can this position be determined. In some cases it is found advantageous to discard the mirror and allow the light from the bull's eye to pass directly into the condenser. This method is especially excellent in photomicrography (see Ch. VIII).

§ 91. **Illuminating the Entire Field.**—With low objectives and large objects, the entire object might not be illuminated if the above method were strictly followed ; in this case, turn the lamp so that the flame is oblique, or if that is not sufficient, continue to turn the lamp until the full width of the flame is used. If necessary the condenser may be lowered, and the concave mirror used. (See also § 80).

REFRACTION AND COLOR IMAGES.

§ 92. **Refraction Images** are those mostly seen in studying microscopic objects. They are the appearances produced by the refraction of the light on entering and on leaving an object. They therefore depend (a) on the form of the object, (b) on the relative refractive powers of object and mounting medium. With such images the diaphragm should not be too large (see § 79).

If the color and refractive index of the object were exactly like the mounting medium it could not be seen. In most cases both refractive index and color differ somewhat, there is then a combination of color and refraction images which is a great advantage. This combination is generally taken advantage of in histology. Fig. 89 is an example of a purely refractive image.

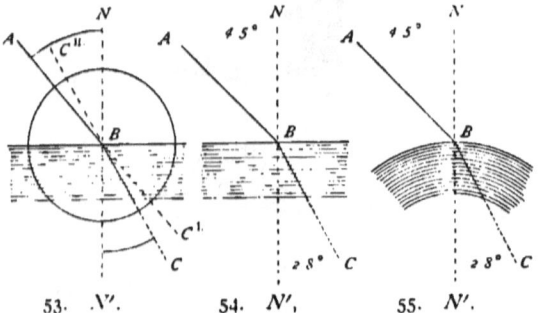

53. N'.   54. N',   55. N'.

FIGS. 53-55.—*Diagrams illustrating refraction in different media and at plane and curved surfaces. In each case the denser medium is represented by line shading and the perpendicular or normal to the refracting surface is represented by the dotted line $N-N'$, the refracted ray by the bent line $A C$.*

¿ 93. **Refraction.**—Lying at the basis of microscopical optics is refraction, which is illustrated by the above figures. It means that light passing from one medium to another is bent in its course. Thus in Fig. 53, light passing from air into water does not continue in a straight line but is bent *toward* the normal $N-N'$, the bending taking place at the point of contact of the air and water; that is, the ray of light $A B$ entering the water at B is bent out of its course, extending to C instead of to C'.

Conversely, if the ray of light is passing from water into air, on reaching the air it is bent *from* the normal, the ray C B passing to A and not in a straight line to C''. By comparing Figs. 54, 55, in which the denser medium is crown glass instead of water, the bending of the rays is seen to be greater as crown glass is denser than water.

It has been found by physicists that there is a constant relation between the angle taken by the ray in the rarer medium, and that taken by the ray in the denser medium. The relationship is expressed thus: Sine of the angle of incidence divided by the sine of the angle of refraction equals the *index of refraction*. In the figures, $\frac{\operatorname{Sin} A B N}{\operatorname{Sin} C B N'}$ index of refraction. Worked out completely in Fig. 53, $A B N = 40°$, $C B N' = 28° 54'$ and $\frac{\operatorname{Sin} 40°}{\operatorname{Sin} 28° 54'} = \frac{0.64279}{0.48327}$ 1.33, *i. e.*, the index of refraction from air to water is 1.33. (See ¿ 30). In Figs. 54-55, illustrating refraction in crown glass, the angles being given, the problem is easily solved as just illustrated. (For table of natural sines see third page of cover; for interpolation see p. 18, ¿ 29a).

¿ 93 a. **Absolute Index of Refraction.**—This is the index of refraction obtained when the incident ray passes from a vacuum into a given medium. As the index of the vacuum is taken as unity, the absolute index of any substance is always greater than unity. For many purposes, as for the purposes of this book, air is treated as if it were a vacuum, and its index is called unity, but in reality the index of refraction of air is about 3 ten-thousandths greater than unity. Whenever the refractive index of a substance is given, the absolute index is meant unless otherwise stated. For example, when the index of refraction of water is said to be 1.33, and of crown glass 1.52, etc., these figures represent the absolute index, and the incident ray is supposed to be in a vacuum.

§ 93 b. **Relative Index of Refraction.**—This is the index of refraction between two contiguous media, as for example between glass and diamond, water and glass, etc. It is obtained by dividing the absolute index of refraction of the substance containing the refracted ray, by the absolute index of the substance transmitting the incident ray. For example, the relative index from water to glass is 1.52 divided by 1.33. If the light passed from glass to water it would be, 1.33 divided by 1.52.

By a study of the figures showing refraction, it will be seen that the greater the refraction the less the angle and consequently the less the sine of the angle, and as the refraction between two media is the ratio of the sines of the angles of incidence and refraction $\left(\frac{\sin i}{\sin r}\right)$ it will be seen that whenever the sine of the angle of refraction is increased, by being in a less refractive medium, the index of refraction will show a corresponding *decrease* and *vice versa*. *That is the ratio of the sines of the angles of incidence and refraction of any two contiguous substances is inversely as the refractive indices of those substances.* The formula is:

$$\left(\frac{\text{Sine of angle of incident ray}}{\text{Sine of angle of refracted ray}}\right) \left(\frac{\text{Index of refraction of refracting medium}}{\text{Index of refraction of incident medium}}\right)$$

Abbreviated $\left(\frac{\sin i}{\sin r}\right) \left(\frac{\text{index } r}{\text{index } i}\right)$. By means of this general formula one can solve any problem in refraction whenever three factors of the problem are known. The universality of the law may be illustrated by the following cases:

(A) Light incident in a vacuum or in air, and entering some denser medium, as water, glass, diamond, etc.

$$\left(\frac{\text{Sin of angle made by the ray in air}}{\text{Sin of angle made by ray in denser medium}}\right) \left(\frac{\text{Index of ref. of denser med.}}{\text{Index of ref. of air (1)}}\right).$$

If the dense substance were glass: $\left(\frac{\sin i}{\sin r}\right) \left(\frac{1.52}{1}\right)$. If the two media were water and glass, the incident light being in water the formula would be: $\left(\frac{\sin i}{\sin r}\right)$ $\left(\frac{1.52}{1.33}\right)$. If the incident ray were in glass and the refracted ray in water: $\left(\frac{\sin i}{\sin r}\right)$ $\left(\frac{1.33}{1.52}\right)$. And similarly for any two media; and as stated above if any three of the factors are given the fourth may readily be found.

§ 93 d. **Critical Angle and Total Reflection.**—In order to understand the Wollaston camera lucida (§ 171, p. 111) and other totally reflecting apparatus, it is necessary briefly to consider the critical angle.

The *critical angle* is the greatest angle that a ray of light in the denser of two contiguous media can make with the normal and still emerge into the less refractive medium. On emerging it will form an angle of 90° with the normal, and if the substances are liquids, the refracted ray will be parallel with the surface of the denser medium.

Total Reflection.—In case the incident ray in the *denser* medium is at an angle with the normal, greater than the *critical angle*, it will be *totally reflected* at the surface of the denser medium, that surface acting as a perfect mirror. By consulting the figures it will be seen that there is no such thing as a critical angle and total reflection in the *rarer* of two contiguous media:

To find the critical angle in the denser of two contiguous media :—
Make the angle of refraction ($i. e.$, the angle in the rarer of the two media) 90° and solve the general equation : $\left(\dfrac{\sin i}{\sin r}\right)$ $\left(\dfrac{\text{index } r}{\text{index } i}\right)$. Let the two substances be water and air, then the sine of r (90°) is 1, the index of air is 1, that of water 1.33, whence $\left(\dfrac{\sin i}{1}\right)$ $\left(\dfrac{1}{1.33}\right)$ or sin i = .751 +. This is the sine of 48°+, and whenever the ray in the water is at an angle of more than 48° it will not emerge into the air, but be totally reflected back into the water.

The case of a ray passing from crown glass into water :
$$\left(\dfrac{\sin i}{\sin r\,(\sin 90°.\ 1)}\right)\ \left(\dfrac{\text{index water }(1.33)}{\text{index glass }(1.52)}\right)\ \text{or}\ \left(\dfrac{\sin i}{1}\right) = \left(\dfrac{1.33}{1.52}\right),$$
whence sin i = .875 sine of critical angle in glass covered with water. The corresponding angle is approximately 61°.

§ 94. **Color Images.**—These are images of objects which are strongly colored, and lighted with so wide an aperture that the refraction images are drowned in the light. Such images are obtained by removing the diaphragm or by using a larger opening. This method of illumination is specially applicable to the study of stained microbes. (See below, § 101).

ADJUSTABLE, WATER AND HOMOGENEOUS OBJECTIVES.
EXPERIMENTS.

§ 95. **Adjustment for Objectives.**—As stated above (§ 22), the aberration produced by the cover-glass (Fig. 56), is compensated for by giving the combinations in the objective a different relative position than they would have if the objective were to be used on uncovered objects. Although this relative position cannot be changed in unadjustable objectives, one can secure the best results of which the objective is capable by selecting covers of the thickness for which the objective was corrected. (See table in § 27). Adjustment may be made also by *increasing* the tube-length for covers *thinner* than the standard and by *shortening* the tube-length for covers *thicker* than the standard (Fig. 57).

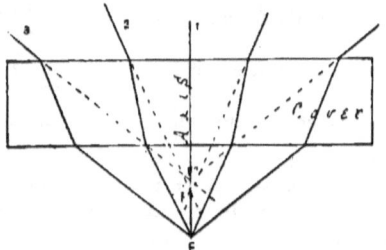

FIG. 56.—*Effect of the cover-glass on the rays from the object to the objective* (*Ross*).

Axis. The projection of the optic axis of the microscope.

F. Focus or axial point of the objective.

F″ and F‴. Points on the axis where rays 2 and 3 appear to originate if traced backward after emerging from the upper side of the cover-glass (*Cover*).

In learning to adjust objectives, it is best for the student to choose some object whose structure is well agreed upon, and then to practice lighting it, shading the stage and adjusting the objective, until the proper appearance is obtained. The adjustment is made by turning a ring or collar which acts on a screw and increases or diminishes the distance between the systems of lenses, usually the front and the back systems (Fig. 40).

General Directions.—(A) The thinner the cover-glass the further must the systems of lenses be separated, $i. e.$, the adjusting collar is turned nearer the zero or the mark "uncovered," and conversely ; (B) the thicker the cover-glass the closer together are the systems brought by turning the adjusting collar *from* the zero mark. This also increases the magnification of the objective (Ch. IV).

The following specific directions for making the cover-glass adjustment are given by Mr. Wenham (Carpenter, 166). "Select any dark speck or opaque portion of the object, and bring the outline into perfect focus ; then lay the finger on the milled-head of the fine motion, and move it briskly backwards and forwards in both directions from the first position. Observe the expansion of the dark outline of the object, both when within and when without the focus. If the greater expansion or coma is when the object is *without* the focus, or farthest from the objective [*i. e.*, in focusing up], the lenses must be placed further asunder, or toward the mark uncovered [*i. e.*, the adjusting collar is turned toward the zero mark as the cover-glass is too thin for the present adjustment]. If the greater expansion is when the object is within the focus, or nearest the objective, [*i. e.*, in focusing down], the lenses must be brought closer together, or toward the mark covered, [*i. e.*, the adjusting collar should be turned away from the zero mark, the cover-glass being too thick for the present adjustment]." *In most objectives the collar is graduated arbitrarily, the zero (O) mark representing the position for uncovered objects. Other objectives have the collar graduated to correspond to the various thickness of cover-glasses for which the objective may be adjusted. This seems to be an admirable plan ; then if one knows the thickness of the cover-glass on the preparation (Ch. VIII) the adjusting collar may be set at a corresponding mark, and one will feel confident that the adjustment will be approximately correct. It is then only necessary for the observer to make the slight adjustment to compensate for the mounting medium or any variation from the standard length of the tube of the microscope.* In adjusting for variations of the length of the tube from the standard it should be remembered that : (**A**) If the tube of the microscope is longer than the standard for which the objective was corrected, the effect is approximately the same as thickening the cover-glass, and therefore the systems of the objective must be brought closer together, *i. e.*, the adjusting collar must be turned *away from* the zero mark. (**B**) If the tube is shorter than the standard for which the objective is corrected, the effect is approximately the same as diminishing the thickness of the cover-glass, and the systems must therefore be separated (Fig. 40).

In using the tube-length for cover correction **Shorten** the tube for too **thick** covers and **Lengthen** the tube for too **thin** covers.

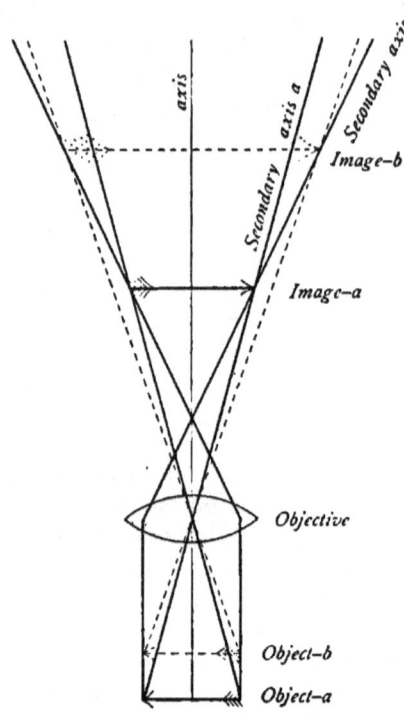

FIG. 57.—*Figure to show that in lengthening the tube of the microscope the object must be brought nearer the principal focus or center of the lens. It will be seen by consulting the figure that in shortening the tube of the microscope the object must be removed farther from the center of the lens. By consulting the figure showing the effect of the cover-glass (Fig. 56) it will be seen that the effect of the cover glass is to bring the object nearer the objective, and the thicker the cover the nearer is the object brought to the objective. As shortening the tube serves to remove the object, it neutralizes the effect of the thick cover, and if the cover is so thin that it does not elevate the object enough for the corrections of the objective, then an increase in the tube-length will correct the defect.*

Furthermore, whatever the interpretation by different opticians of what should be included in "tube-length," and the exact length in millimeters, its importance is very great; for each objective gives the most perfect image of which it is capable with the "tube-length" for which it is corrected, and the more perfect the objective the greater the ill-effects on the image of varying the "tube-length" from this standard. The plan of designating exactly what is meant by "tube-length," and engraving on each objective the "tube-length" for which it is corrected, is to be commended, for it is manifestly difficult for each worker with the microscope to find out for himself for what "tube-length" each of his objectives was corrected. (See appendix).

§ 97. **Water Immersion Objectives.**—Put a water immersion objective in position (§ 43) and the fly's wing for object under the microscope. Place a drop of distilled water on the cover-glass, and with the coarse adjustment lower the tube till the objective dips into the water, then light the field well and turn the fine adjustment one way and the other till the image is clear. Water immersions are exceedingly convenient in studying the circulation of the blood, and for many other purposes where aqueous liquids are liable to get on the cover-glass. If the objective is adjustable, follow the directions given in § 95.

When one is through using a water immersion objective, remove it from the microscope and with some lens paper wipe all the water from the front-lens. Unless this is done dust collects and sooner or later the front-lens would be clouded. It is better to use distilled water to avoid the gritty substances that are liable to be present in natural waters, as these gritty particles might scratch the front-lens.

HOMOGENEOUS IMMERSION OBJECTIVES : EXPERIMENTS.

§ 98. As stated above, these are objectives in which a liquid of the same refractive index as the front-lens of the objective is placed between the front-lens and the cover-glass.

§ 99. **Tester for Homogeneous Liquid.**—In order that full advantage be derived from the homogeneous immersion principle, the liquid employed must be truly homogeneous. To be sure that such is the case, one may use a tester like that constructed by the Gundlach Optical Co., then if the liquid is too dense it may be properly diluted and *vice versa*. For the cedar oil immersion liquid, the density may be diminished by the addition of pure cedar wood oil. The density may be increased by allowing it to thicken by evaporation. (See H. L. Smith, Proc. Amer. Soc. Micr., 1885, p. 83, and appendix).

§ 100. **Refraction Images.**—Put a 2 mm. ($\frac{1}{12}$th in.) homogeneous immersion objective in position, employ an illuminator. Use some histological specimen like a muscular fiber as object, make the diaphragm opening about 3 mm. in diameter, add a drop of the homogeneous immersion liquid and focus as directed in § 70. The object will be clearly seen in all details by the unequal refraction of the light traversing it. The difference in color between it and the surrounding medium will also increase the sharpness of the outline. If an air bubble preparation (§ 73) were used, one would get pure, refraction images.

§ 101. **Color Images.**—Use some stained microbes, as *Bacillus tuberculosis* for object. Put a drop of the immersion liquid on the coverglass or the front lens of the homogeneous objective. Remove the diaphragms from the illuminator or in case the iris diaphragm is used, open to its greatest extent. Focus the objective down so that the immersion fluid is in contact with both the front lens and the cover-glass, then with the fine adjustment get the microbes in focus. They will stand out as clearly defined colored objects on a bright field.

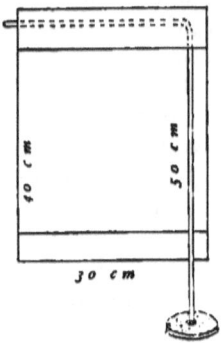

FIG. 58.—*Screen for shading the microscope and the face of the observer. This is very readily constructed as shown in the figure by supporting a wire in a disc of lead, iron, or heavy wood. The screen is then completed by hanging over the bent wire, cloth or manilla paper 30 x 40 cm. The lower edge of the screen should be a little below the stage of the microscope and the upper edge high enough to screen the eyes of the observer.*

§ 102. **Shading the Object.**—To get the clearest image of an object no light should reach the eye except from the object. A handkerchief or a dark cloth wound around the objective will serve the purpose. Often the proper effect may be obtained by simply shading the top of the stage with the hand or with a piece of bristol board. Unless one has a very favorable light the shading of the object is of the greatest advantage, especially with homogeneous immersion objectives. The screen (Fig. 58) is the most satisfactory means for this purpose, as the entire microscope above the illuminating apparatus is shaded.

§ 103. **Cleaning Homogeneous Objectives.**—After one is through with a homogeneous objective, it should be carefully cleaned as follows: Wipe off the homogeneous liquid with a piece of the lens paper (§ 107), then if the fluid is cedar oil, wet one corner of a fresh piece in benzin and wipe the front lens with it. Immediately afterward wipe with a dry part of the paper. The cover-glass of the preparation can be cleaned in the same way. If the homogeneous liquid is a glycerin mixture proceed as above, but use water instead of benzin to remove the last traces of glycerin.

CARE OF THE MICROSCOPE.

§ 104. The microscope should be handled carefully and kept perfectly clean. The oculars and objectives should never be allowed to fall. When not in use keep it in a place as free as possible from dust.

All parts of the microscope should be kept free from liquids, especially from acids, alkalies, alcohol, benzin, turpentine and chloroform.

§ 105. **Care of the Mechanical Parts.**—To clean the mechanical parts put a small quantity of some fine oil (olive oil or liquid vaselin and benzin, equal parts), on a piece of chamois leather or on the lens paper, and rub the parts well, then with a clean dry piece of the chamois or paper wipe off most of the oil. If the mechanical parts are kept clean in this way a lubricator is rarely needed. Where opposed brass surfaces "cut," *i. e.*, when from the introduction of some gritty material, minute grooves are worn in the opposing surfaces, giving a harsh movement, the opposing parts should be separated, carefully cleaned as described above and any ridges or prominences scraped down with a knife. Where the tendency to "cut" is marked, a very slight application of equal parts of beeswax and tallow, well melted together, serves a good purpose.

In cleaning lacquered parts, benzin alone answers well, but it should be quickly wiped off with a clean piece of the lens paper. Do not use alcohol as it dissolves the lacquer.

§ 106. **Care of the Optical Parts.**—These must be kept scrupulously clean in order that the best results may be obtained.

Glass surfaces should never be touched with the fingers, for that will soil them.

The glass of which the lenses are made is quite soft, consequently it is necessary that only soft, clean cloths or paper be used in wiping them.

Whenever an objective is left in position on a microscope, or when several are attached by means of a revolving nose-piece, an ocular should be left in the upper end of the tube to prevent dust from falling down upon the back-lens of the objective.

§ 107. **Lens Paper.**—The so-called Japanese filter paper, which from its use with the microscope, I have designated lens paper, has been used in the author's laboratory for the last ten years for cleaning the lenses of oculars and objectives, and especially for removing the fluid used with immersion objectives. Whenever a piece is used once it is thrown away. It has proved more satisfactory than cloth or chamois, because dust and sand are not present ; and from its bibulous character it is very efficient in removing liquid or semi-liquid substances.

§ 108. *Dust* may be removed with a camel's hair brush, or by wiping with the lens paper.

Cloudiness may be removed from the glass surfaces by breathing on them, then wiping quickly with a soft cloth or the lens paper.

Cloudiness on the inner surfaces of the ocular lenses may be removed by unscrewing them and wiping as directed above. A high objective should never be taken apart by an inexperienced person.

If the cloudiness cannot be removed as directed above, moisten one corner of the cloth or paper with 95 per cent. alcohol, wipe the glass first with this, then with the dry cloth or the paper.

Water may be removed with soft cloth or the paper.

Glycerin may be removed with cloth or paper saturated with distilled water; remove the water as above.

Blood or other albuminous material may be removed while fresh with a moist cloth or paper, the same as glycerin. If the material has dried to the glass, it may be removed more readily by adding a small quantity of ammonia to the water in which the cloth is moistened, (water 100 cc., ammonia 1 cc).

Canada Balsam, damar, paraffin, or any oily substance, may be removed with a cloth or paper wet with chloroform, benzin or xylene. The application of these liquids and their removal with a soft, dry cloth or paper should be as rapid as possible, so that none of the liquid will have time to soften the setting of the lenses.

Shellac Cement may be removed by the paper or a cloth moistened in 95 per cent. alcohol.

Brunswick Black, Gold Size, and all other substances soluble in chloroform, etc., may be removed as directed for balsam and damar.

In general, use a solvent of the substance on the glass and wipe it off quickly with a fresh piece of the lens paper.

It frequently happens that the upper surface of the back combination of the objective becomes dusty. This may be removed in part by a brush, but more satisfactorily by using a piece of the soft paper loosely twisted. When most of the dust is removed some of the paper may be put over the end of a pine stick (like a match stick) and the glass surface carefully wiped.

CARE OF THE EYES.

§ 109. Keep both eyes open, using the eye-screen if necessary (Fig. 59, 60); and divide the labor between the two eyes, *i. e.*, use one eye for observing the image awhile and then the other. In the beginning

LIGHTING AND FOCUSING.

it is not advisable to look into the microscope continuously for more than half an hour at a time. One never should work with the microscope after the eyes feel fatigued. After one becomes accustomed to microscopic observation he can work for several hours with the microscope without fatiguing the eyes. This is due to the fact that the eyes become inured to labor like the other organs of the body by judicious exercise. It is also due to the fact that but very slight accommodation is required of the eyes, the eyes remaining nearly in a condition of rest as for distant objects. The fatigue incident upon using the microscope at first is due partly at least to the constant effort on the part of the observer to remedy the defects of focusing of the microscope by accommodation of the eyes. This should be avoided and the fine adjustment of the microscope used instead of the muscles of accommodation. With a microscope of the best quality, and suitable light—that is light which is steady and not so bright as to dazzle the eyes nor so dim as to strain them in determining details—microscopic work should improve rather than injure the sight.

FIG. 59.—*Ward's Eye-Shade.*

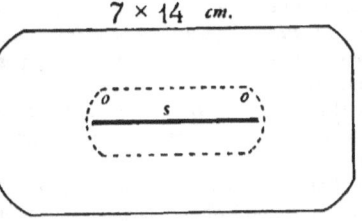

FIG. 60. *Double Eye-Shade. This is readily made by taking some thick bristol board 7 x 14 centimeters and making an oblong opening with rounded ends (o—o) and of such a diameter that it goes readily over the tube of the microscope. This is then covered on both sides with velveteen and a central slit (s) made in the cloth. This admits the tube of the microscope and holds the screen in position. It may readily be pulled from side to side and thus serves for either eye, or for the use of the eyes alternately.*

§ 110. **Position and Character of the Work-Table.**—The work-table should be very firm and large (122 x 72 cm.; 28 x 48 in.), so that the necessary apparatus and material for work may not be too crowded. The table should also be of the right height to make work by it comfortable. An adjustable stool, something like a piano stool is convenient, then one may vary the height corresponding to the necessities of special cases. It is a great advantage to sit facing the window if daylight is used, then the hands do not constantly interfere with the illumination. To avoid the discomfort of facing the light a screen like that shown in Fig. 58 is very useful (see also under lighting, § 58).

TESTING THE MICROSCOPE.

§ 111. **Testing the Microscope.**—To be of real value this must be accomplished by a person with both theoretical and practical knowledge, and also with an unprejudiced mind. Such a person is not common, and when found, does not show an over anxiety to pass judgment. Those most ready to offer advice should as a rule be avoided, for in most cases they simply "have an ax to grind," and are sure to commend only those instruments that conform to the "fad" of the day. From the writer's experience it seems safe to say that the inexperienced can do no better than to trust to the judgment of one of the optical companies. The makers of microscopes and objectives guard with jealous care the excellence of both the mechanical and optical part of their work, and send out only instruments that have been carefully tested and found to conform to the standard. This would be done as a matter of business prudence on their part, but it is believed by the writer that microscope makers are artists first and take an artist's pride in their work, they therefore have a stimulus to excellence greater than business prudence alone could give.

§ 112. **Mechanical Parts.**—All of the parts should be firm, and not too easily shaken. Bearings should work smoothly. The mirror should remain in any position in which it is placed.

Focusing Adjustments.—The coarse or rapid adjustment should be by rack and pinion, and work so smoothly that even the highest power can be easily focused with it. In no case should it work so easily that the body of the microscope is liable to run down and plunge the objective into the object. If any of the above defects appear in a microscope that has been used for some time, a person with moderate mechanical instinct will be able to tighten the proper screws, etc.

The Fine Adjustment is more difficult to deal with. From the nature of its purpose, unless it is approximately perfect, it would better be off the microscope entirely.

It should work smoothly and be so balanced that one cannot tell by the feeling when using it whether the screw is going up or down. Then there should be absolutely no motion except in the direction of the optic axis, otherwise the image will appear to sway even with central light. Compare the appearance when using the coarse and when using the fine adjustment. There should be no swaying of the image with either if the light is central (§ 73).

§ 113. **Testing the Optical Parts.**—As stated in the beginning, this can be done satisfactorily only by an expert judge. It would be of very great advantage to the student if he could have the help of such a person. In no case is the condemnation of a microscope to be made by an inexperienced person. If the beginner will bear in mind that his failures are due mostly to his own lack of knowledge and lack of skill; and will truly endeavor to learn and apply the principles laid down in this and in the standard works referred to, he will learn after a while to estimate at their true value all the pieces of his microscope. (See appendix).

LABORATORY COMPOUND MICROSCOPES.

§ 114. Optical Parts.—A great deal of beginning work with the microscope in biological laboratories is done with simple and inexpensive apparatus. Indeed if one contemplates the large classes in the universities and medical schools, it can be readily understood that microscopes costing from \$25-50 each and magnifying from 25 to 500 diameters, are all that can be expected. But for the purpose of modern histological investigation and of advanced microscopical work in general, a microscope should have something like the following character: Its optical outfit should comprise, (a) dry objectives of 50 mm. (2 in.), 16-18 mm. ($\frac{1}{4}$ in.) and 3 mm. ($\frac{1}{8}$ in.) equivalent focus. There should be present also a 2 mm. ($\frac{1}{12}$ in.) or 1.5 mm. ($\frac{1}{16}$ in.) homogeneous immersion objective. Of oculars there should be several of different power. An illuminator or substage condenser, and an Abbe camera lucida are also necessities, and a micro-spectroscope and a micro-polarizer are very desirable.

Even in case all the optical parts cannot be obtained in the beginning, it is wise to secure a stand upon which all may be used when they are finally secured.

As to the objectives. The best that can be afforded should be obtained. Certainly at the present, the apochromatics stand at the head, although the best achromatic objectives approach them very closely.

§ 115. Mechanical Parts or Stand.—The stand should be low enough so that it can be used in a vertical position on an ordinary table without inconvenience; it should have a jointed (flexible) pillar for inclination at any angle to the horizontal. The adjustments for focusing should be two,—a coarse adjustment or rapid movement with rack and pinion, and a fine adjustment by means of a micrometer screw. Both adjustments should move the entire tube of the microscope. The body or tube should be short enough for objectives corrected for the short or 160 millimeter tube-length, and the draw-tube should be graduated in centimeters and millimeters. The lower end of the draw tube and of the tube should each possess a standard screw for objectives (frontispiece). The stage should be quite large for the examination of slides with serial sections. If there is no mechanical stage (§ 116), it is also of considerable advantage to have the stage with a circular, revolving top, and two centering screws with milled heads. In this way a mechanical stage with limited motion is secured, and this is of the highest advantage in using powerful objectives. The sub-stage fittings should be so arranged as to enable one to dispense entirely with diaphragms, to use ordinary diaphragms, or to use the condenser. The condenser mounting should allow up and down motion, preferably by rack and pinion. The base should be sufficiently heavy and so arranged that the microscope will be steady in all positions, and interfere the least possible amount with the manipulation of the mirror and other sub-stage accessories.

§ 116. Mechanical Stage.—There should also be present some form of mechanical stage. That on the most expensive American and English microscopes for the last twenty years and the one now present on the larger continental microscopes, is excellent for high powers and preparations of moderate dimensions, but for the study of serial sections and large sections or preparations in general, mechanical stages like those shown in Figs. 68-69 are more useful. This form of mechanical stage has the advantage of giving great lateral and forward and backward motion. It is a modification of the mechanical stage of Tolles. The modification consists in removing the thin plate and instead, having a clamp to catch the ends of the glass slide. The slide is then moved on the face of the stage proper. This modification was first made by Mayall. It has since been modified by Reichert, Zeiss, Leitz and others in Europe and by the Bausch & Lomb Optical Co. in America.— (Jour. Roy. Micr. Soc., 1885, p. 122. See also Zeit. wiss. Mikroskopie, (II), 1885, pp. 289-295 ; 1887, (IV,, pp. 25-30).

§ 117. **Society Screw.**—Owing to the lack of uniformity in screws for microscopic objectives, the Royal Microscopical Society of London, in 1857, made an earnest effort to introduce a standard size. The specifications of this standard are as follows : " Whitworth thread, $i\ e.$, a V shaped thread, sides of thread inclined to an angle of $55°$ to each other, one-sixth of the V depth of the thread being rounded off at the top of the thread, and one-sixth of the thread being rounded off at the bottom of the thread. Pitch of screw, 36 to the inch ; length of thread on object-glass, 0.125 inch ; plain fitting above thread of object glass, 0.15 inch long, to be about the size of the bottom of male thread ; length of thread of nose-piece [on the lower end of the tube of the microscope], not less than 0.125 inch ; diameter of the object-glass screw at the bottom of the screw, 0.7626 inch ; diameter of the nose-piece screw at the bottom of the thread, 0.8 inch.

In order to facilitate the introduction of this universal screw, or as it soon came to be called " *The Society Screw*," the Royal Microscopical Society undertook to supply standard taps. From the mechanical difficulty in making these taps perfect there soon came to be considerable difference in the " Society Screws," and the object of the society in providing a universal screw was partly defeated.

In 1884 the American Microscopical Society appointed Mr. Edward Bausch and Prof. William A Rogers upon a committee to correspond with the Royal Microscopical Society, with a view to perfecting the standard " Society Screw," or of adopting another standard and of perfecting methods by which the screws of all makers might be truly uniform. Although this matter was earnestly considered at the time by the Royal Microscopical Society, the mechanical difficulties were so great that the improvements were abandoned.

Fortunately, however, during the present year (1896) that society has again taken hold of the matter in earnest, and we are now promised a new " Society Screw " which shall be accurate ; and facilities for obtaining the standard will be so good that there is a reasonable certainty that the universal screw for microscopic objectives may be realized. It is indeed astonishing to see how widely spread the " Society Screw " has become. Indeed there is not a maker of first class microscopes in the world who does not supply the objectives and stands with the " Society Screw," and an objective in England or America which does not have this screw should be looked upon with suspicion. That is, it is either old, cheap, or not the product of one of the great opticians. For the Standard, or "Society Screw," see : Trans. Roy. Micr. Soc., 1857, pp. 39-41 ; 1859, pp. 92-97 ; 1860, pp. 103-104. (All to be found in Quar. Jour. Micr. Sci., o. s., vols. VI, VII and VIII). Proc. Amer. Micr. Soc., 1884, p. 274 ; 1886, p. 199 ; 1893, p. 38. Journal of the Royal Microscopical Society, Aug., 1896.

In this last paper of four pages the matter is very carefully gone over and full specifications of the new screw given. It conforms almost exactly with the original standard adopted by the society, but means have been devised by which it may be kept standard.

FIGURES OF LABORATORY MICROSCOPES AND ACCESSORY APPARATUS.

It was deemed advisable in this new edition to figure some of the most common of the laboratory microscopes and this has been rendered possible mostly by the courtesy of the makers and importers. During the last five years very great vigor has been shown in the microscopical world. This has been stimulated largely by the activity in biological science and the widespread appreciation of the microscope, not only as a desirable, but as a necessary instrument of study and research. The production of the new kinds of glass (Jena glass), and the apochromatic objectives have been a no less potent factor in promoting progress. It is gratifying also to know that with the increase in the use of the microscope, not only are the optical and mechanical parts improved, and that very greatly, but the price has decreased so that at the present time schools cannot afford to be without one or more, and individuals are not debarred from the possession of an instrument adequate to their needs. The cost of a complete outfit varies from 25 to 600 dollars. The student is advised to write to one or more of the opticians for complete catalogs. See list, p. 2 of cover.

§ 118. **Marker for Preparations.** (Figs. 61-66).—This instrument consists of an objective-like attachment which may be screwed into the nose-piece of the microscope. It bears on its lower end (Figs. 61-3) a small brush and the brush can be made more or less eccentric and can be rotated, thus making a larger or smaller circle. In using the marker the brush is dipped in colored shellac or other cement and when the part of the preparation to be marked is found and put exactly in the middle of the field the objective is turned aside and the marker turned into position. The brush is brought carefully in contact with the cover-glass and rotated. This will make a delicate ring of the colored cement around the object. Within this very small area the desired object can be easily found on any microscope. The brush of the marker should be cleaned with 95 % alcohol after it is used. (Proc. Amer. Micr. Soc., 1894, pp. 112-118).

§ 119. **Pointer in the Ocular.**—The Germans have a pointer ocular (Spitzen-Okular), an ocular with one or two delicate rods or pointers at the level of the real image, that is, at the level of the diaphragm (Figs. 21, 30 D). For the purposes of demonstrating any particular structure or object in the field, a temporary pointer may be easily inserted in any ocular as follows: Remove the eye-lens and with a little mucilage or Canada balsam fasten an eye lash (cilium) to the diaphragm (Fig. 30 D) so that it will project about half way across the opening. If one uses this ocular, the pointer will appear in the field and one can place the specimen so that the pointer indicates it exactly, as in using a pointer on a diagram or on the black-board. It is not known to the author who devised this method. It is certainly of the greatest advantage in demonstrating objects like amoebas or white blood corpuscles to persons not familiar with them, as the field is liable to have in it many other objects which are more easily seen, as the red blood corpuscles or particles of vegetation or dirt in the case of the blood preparation or of the amoeba.

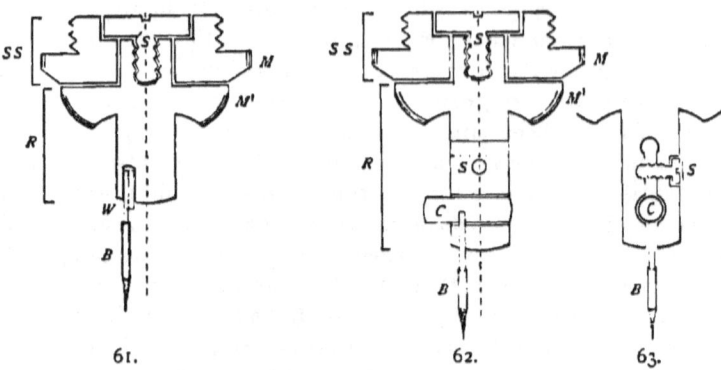

61.   62.   63.

FIGS. 61-63. *Sectional Views of the two Forms of the Marker.*

FIG. 61. *The simplest form of marker. It consists of the part SS with the milled edge (M). This part bears the Society or objective screw for attaching the marker to the microscope. R. Rotating part of the marker. This bears the eccentric brush (B) at its lower end. The brush is on a wire (W). This wire is eccentric, and may be made more or less so by bending the wire. The central dotted line coincides with the axis of the microscope. The revolving part is connected with the " Society Screw " by the small screw (S).*

FIG. 62. *SS, R, and B. All parts same as with Fig. 61, except that the brush is carried by a sliding cylinder the end view being indicated in Fig. 63.*

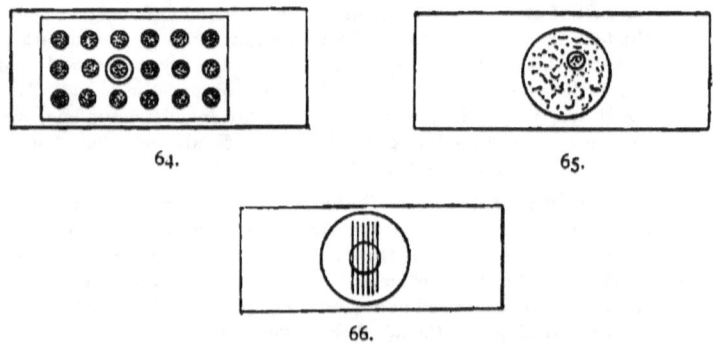

64.   65.

66.

FIGS. 64, 65, 66. *Specimens Showing the Use of the Marker.*

In Fig. 64 a section of a series is marked to indicate that this section shows something especially well. In Fig. 65 some blood corpuscles showing ingested carbon very satisfactorily are surrounded by a minute ring, and in Fig. 66 the lines of a micrometer are ringed to facilitate finding the lines.

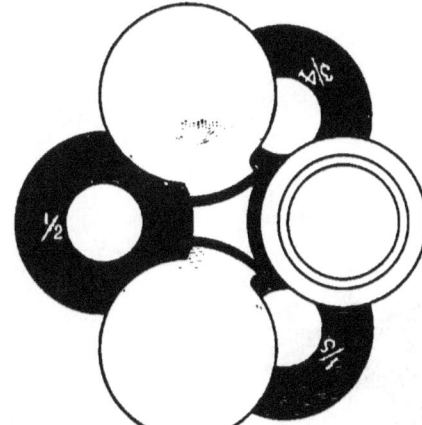

FIG. 67. *Krauss' Method of Marking Objectives on a Revolving Nose-Piece.*

As seen in the figure, the equivalent focus of the objective is engraved on the diaphragm above the back lens and may be very readily seen in rotating the nose-piece. This is of great advantage and facilitates the changing of objectives, as one can see what objective is coming into place without trouble.

FIGS. 68 69. *The Tolles-Mayall Mechanical Stage* (⅔ 116).

Both these mechanical stages have the great advantage of large movement in both directions, so that a series may be studied with great certainty and facility. Both have scales and verniers, so that the position of any particular feature of a preparation may be readily refound. The figures on the scale being different there is never doubt as to the position of each from the record.

FIG. 68. *The Tolles Mayall mechanical stage as constructed by Leitz.* It is shown in position on the stage of the microscope; it is fastened to the stage by a pin and screw near the pillar.

FIG. 69. *The Tolles Mayall mechanical stage made by the Bausch & Lomb Optical Company.* It is separated from the microscope. It is attached to the microscope by a clamp surrounding the pillar. This form of connection was employed by Reichert & Zeiss in the earlier forms devised, and is still used by them

5

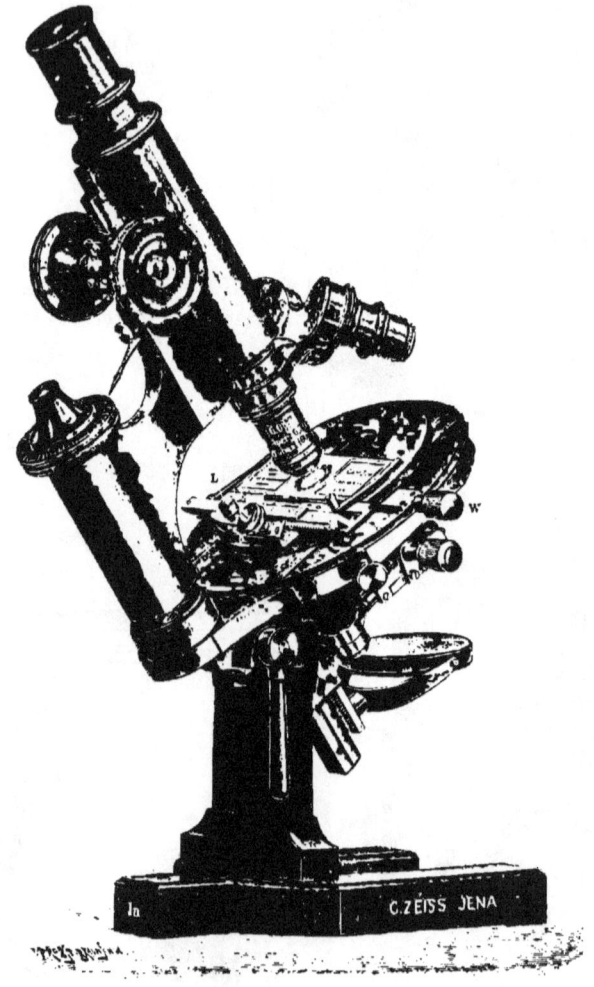

FIG. 70. *Zeiss' Microscope I^a with Mechanical Stage. This figure, from Zeiss' Catalog No. 30, represents the Continental Model of Microscope in its most perfect form.*

K. Milled head of the screw for the lateral movements of the stage.

L. Screw for fixing the laterally moving mechanism of the stage. By unscrewing this the laterally moving part may be removed, leaving a plain stage.

W. Screw for moving the stage forward and backward.

CH. II.] LABORATORY MICROSCOPES. 65 a

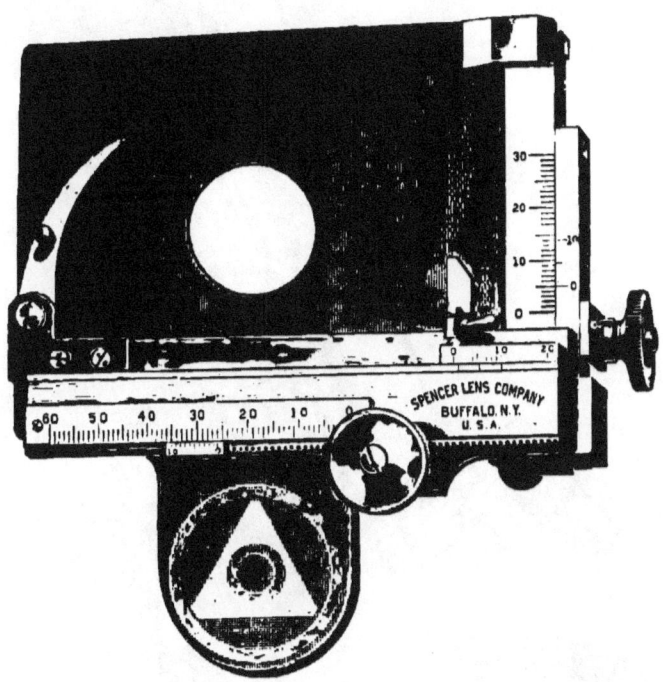

FIG. 70a.—*Spencer-Winkel form of Mechanical Stage. It is readily attachable to any microscope with rectangular stage, by two clamping screws on the right. Its range of motion is about 30 × 60 mm., thus enabling one to study as large a series as is usually put on an ordinary slide.*

FIG. 70b.—(*See next page.*) *Spencer Lens Company's Microscope No. 1. This microscope is also furnished with a tube and oculars of the same diameter as those on the Zeiss microscopes for investigators who prefer the small tube.*

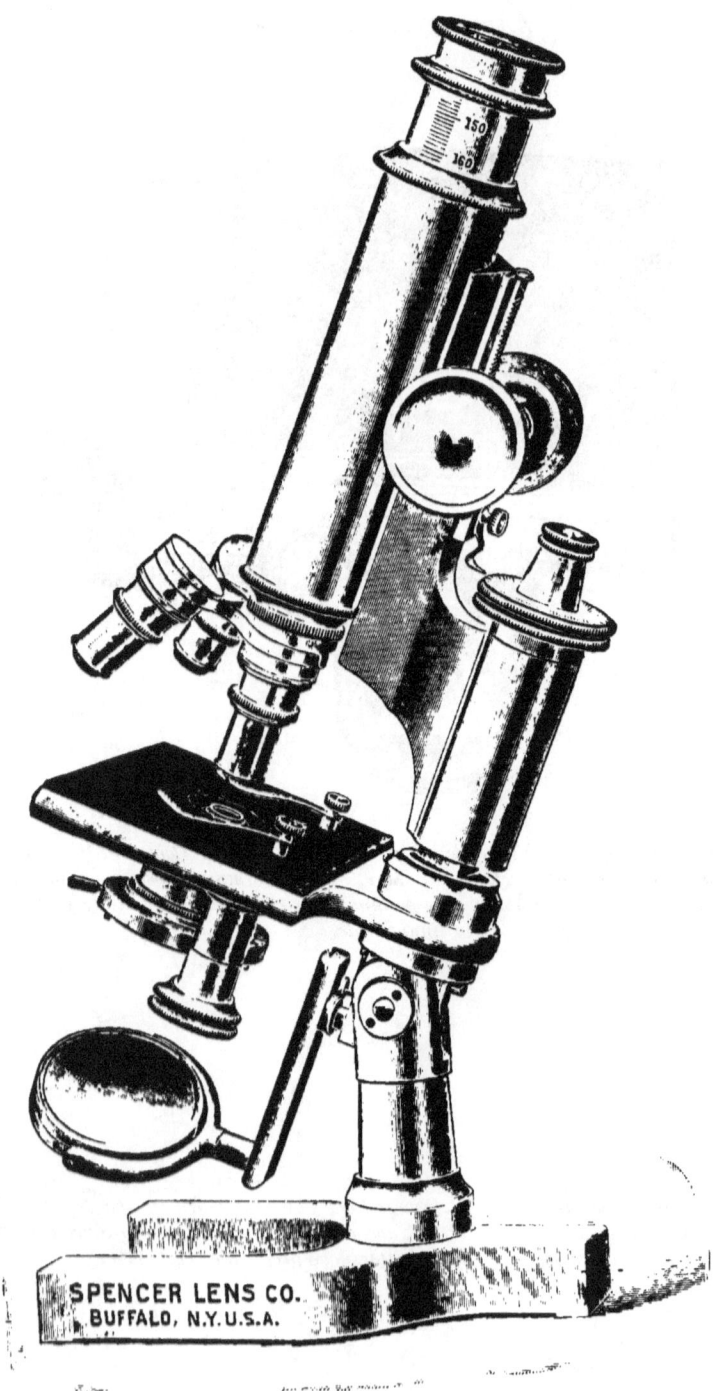

LABORATORY MICROSCOPES.

FIG. 71. *Watson & Sons' Edinburgh Students' Microscope (Stand G). This is a good representative of the tripod-base, English models of to-day. It is in general like the Powell and Lealand stands which have held their position with the foremost English microscopists for the last 40 years. (See Carpenter-Dallinger, p. 172).*

Microscopes with tripod bases something after this pattern are now being made on the Continent.

Attention is called to the sub-stage for the condenser. It possesses centering screws so that any apparatus used in it may be accurately centered. It is to be hoped that all microscopes of this grade will soon be supplied with a centering sub-stage.

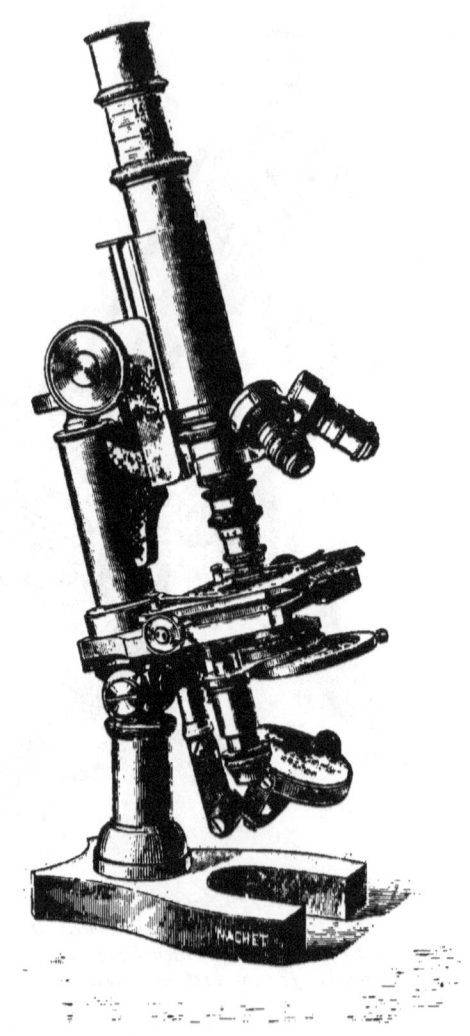

FIG. 72. *Natchet & Fils' Medium Microscope No. 4 with Movable Stage.* The French microscopes set the fashion for the Continent of Europe, and the Continental or *Hartnack Model* with the horse shoe base has extended to all lands, but see *Carpenter-Dallinger*, p. 208.

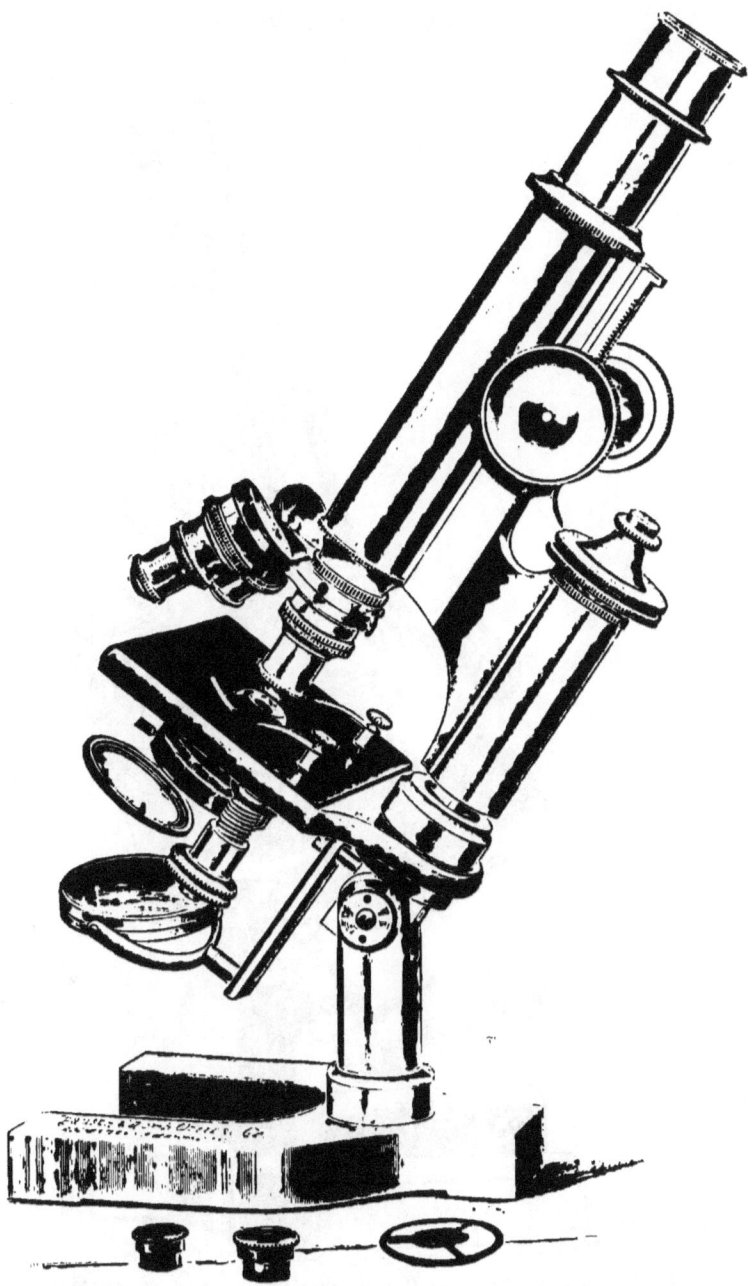

FIG. 73. *The BB Microscope of the Bausch & Lomb Optical Co.*

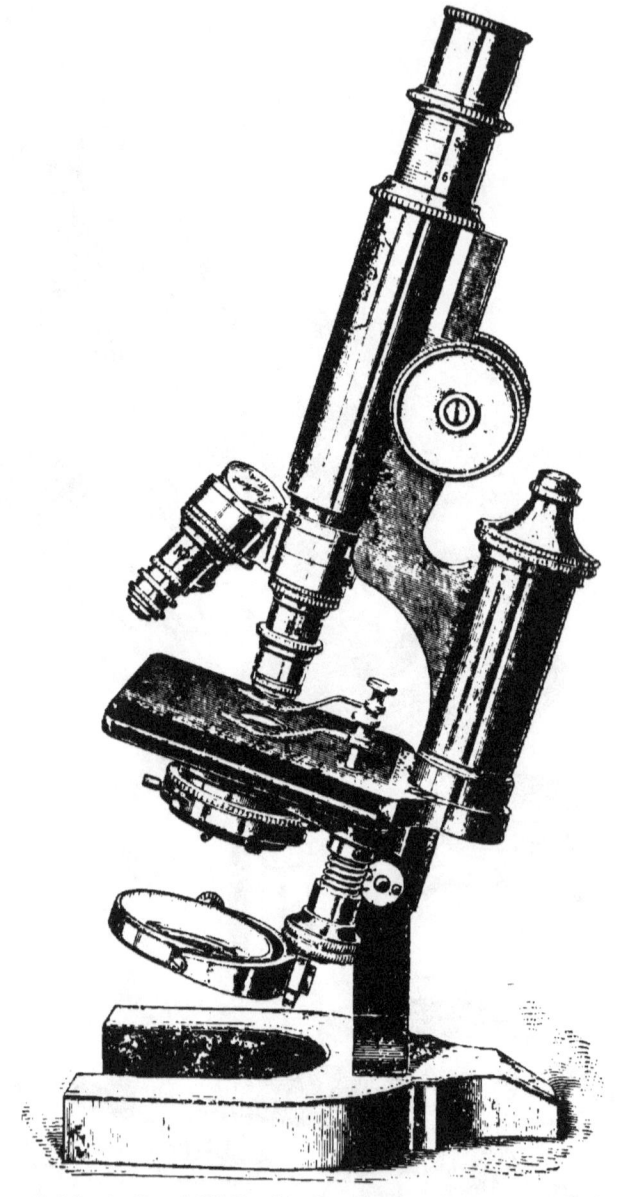

FIG. 74. *Reichert's Stand III B, after Specifications by Richards & Co*

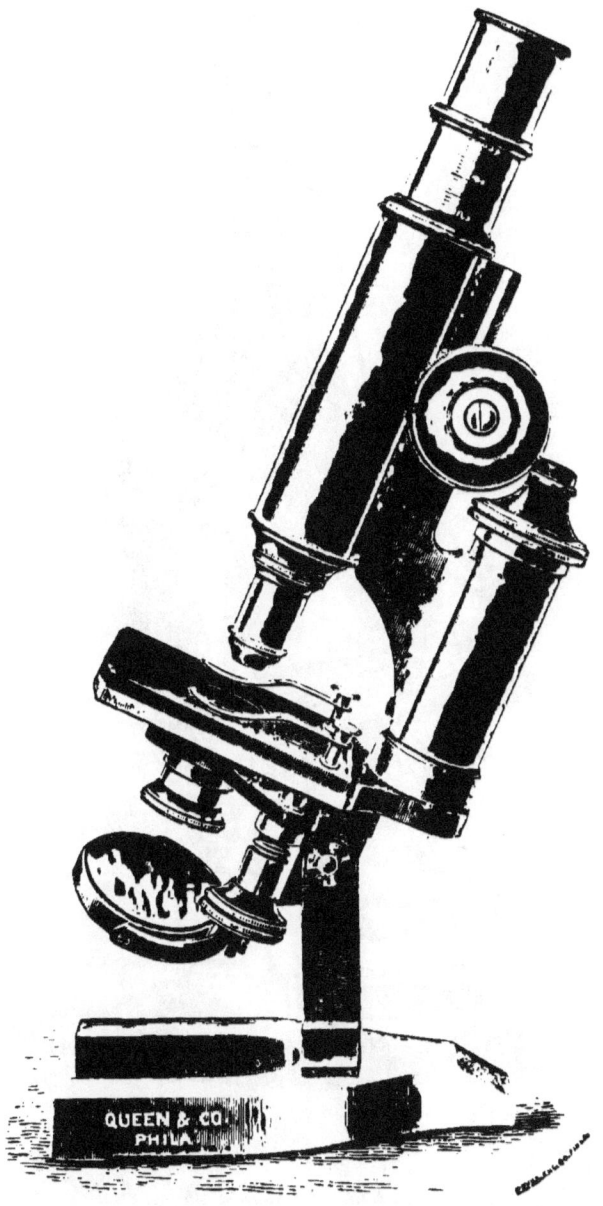

FIG. 75. *Queen & Company's Microscope of the Continental Pattern, No. II.*

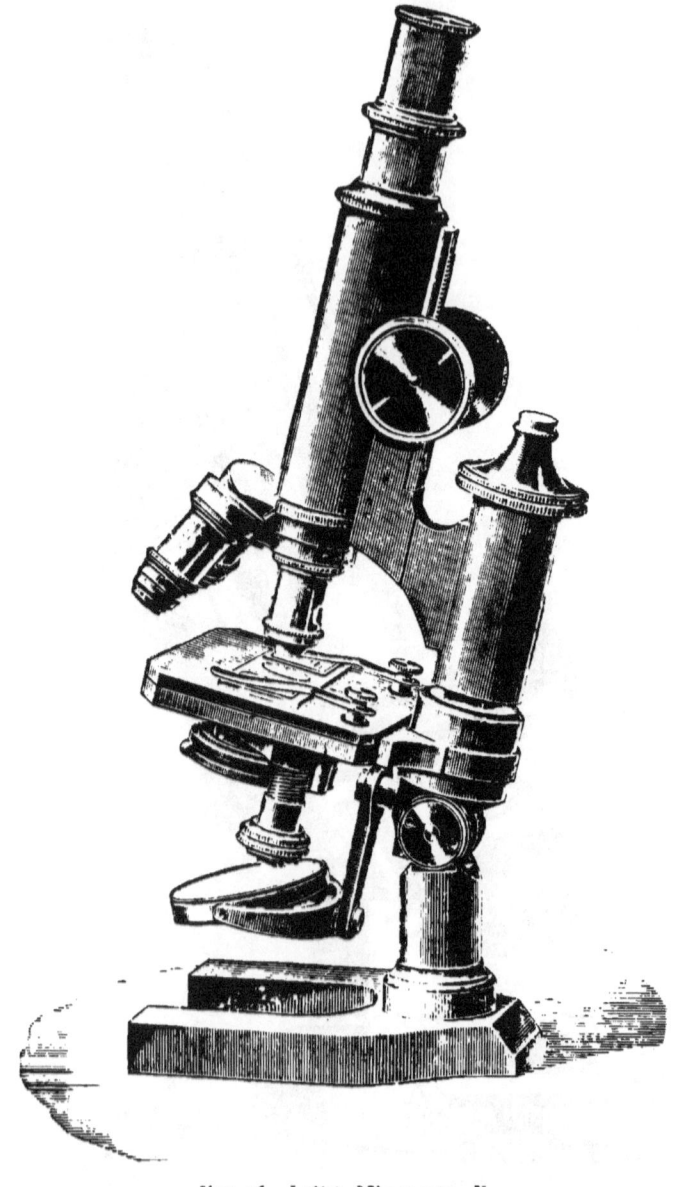

FIG. 76. *Leitz' Microscope, Ib.*

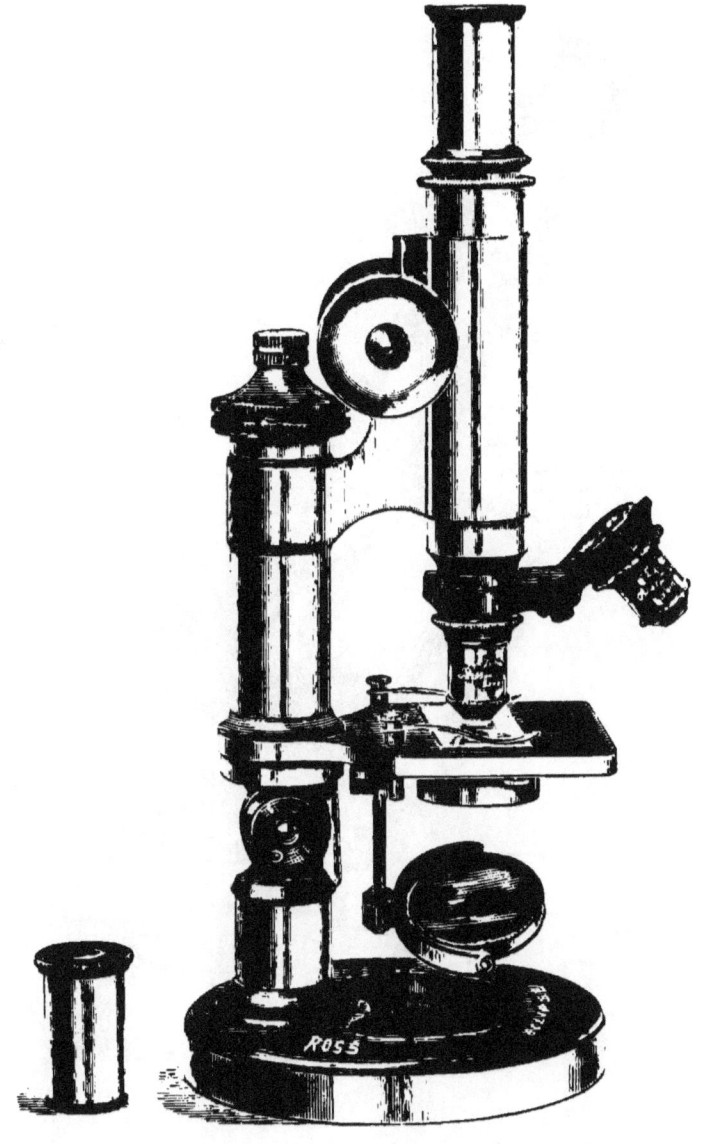

FIG. 77. *Ross Eclipse Microscope. The heavy circular base rotates to give firmness upon inclination.*

74 LABORATORY MICROSCOPES. [CH. II.

FIG. 78. *The AA Microscope of the Bausch & Lomb Optical Co., with sliding tube instead of rack and pinion. This microscope is also furnished with rack and pinion.*

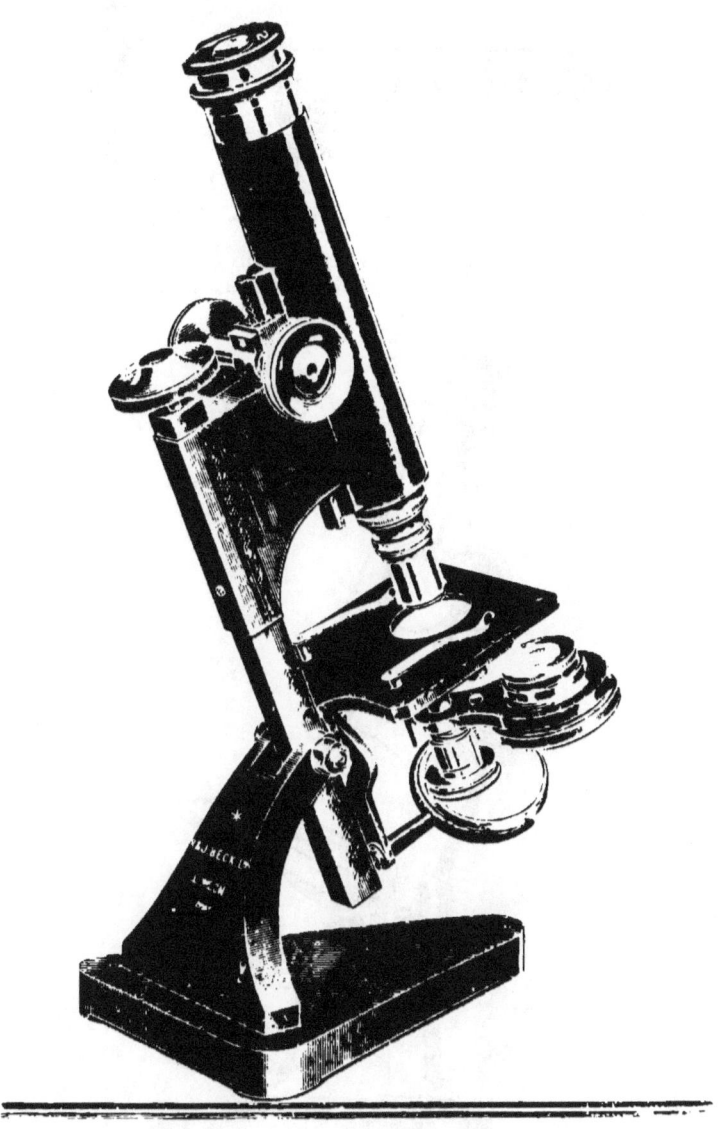

FIG. 79. *Beck's Star Microscope.*

FIG. 80. *Zentmayer's Clinical Microscope.* This is supplied with a lamp and so mounted that it may readily be passed around a class for the purpose of demonstrating some microscopic structure.

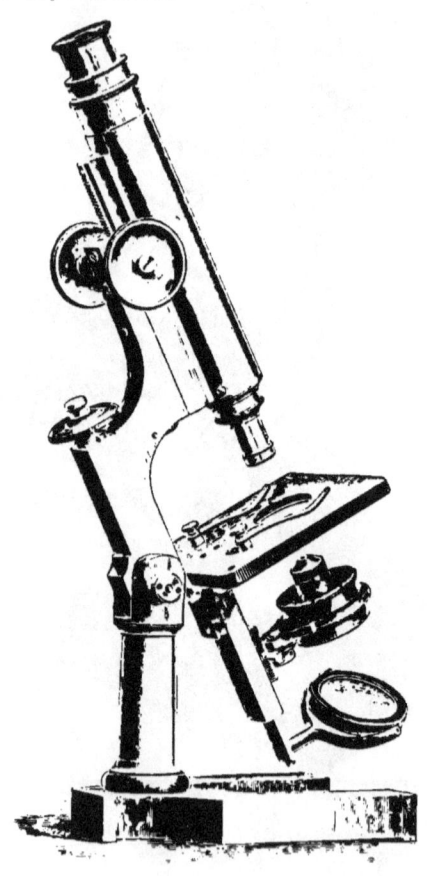

FIG. 81. *Zentmayer's Microscope, No. V.*

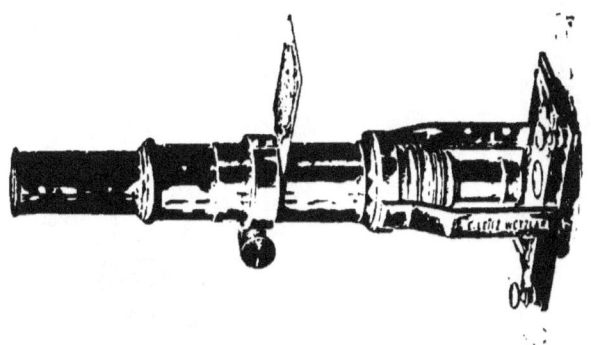

FIG. 82. *Leitz' Demonstration Microscope.* This is designed for class demonstration and is pointed toward the window, or some other source of light, by the student. It has an arrangement for holding a sketch of the object under the microscope.

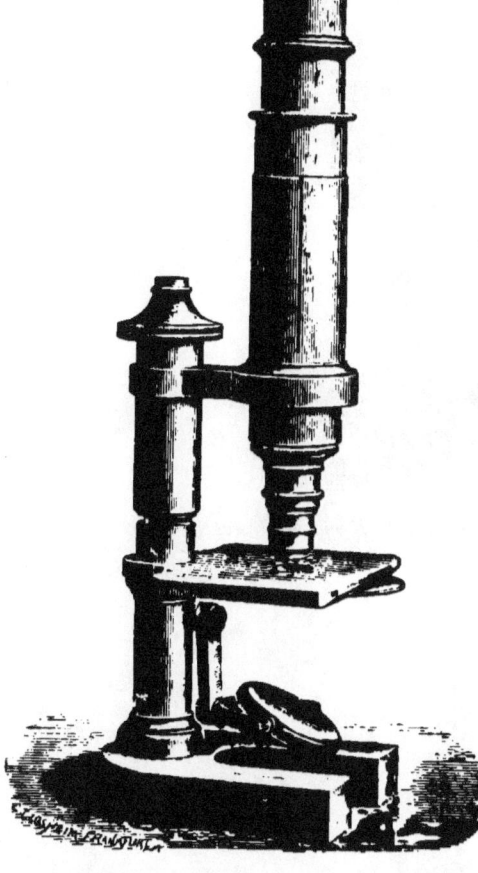

FIG. 83. *Leitz' Microscope, No. IV*, with sliding tube for coarse adjustment, and no joint for inclination. This microscope must always be used in the vertical position. Microscopes of this form are not so much the fashion as they were a few years ago. If any instrument requires mechanical aids to enable one to attain a desired result with ease and certainty, it is a microscope.

Most of the Leitz microscopes are now supplied with joint and with rack and pinion for the coarse adjustment (Fig. 76).

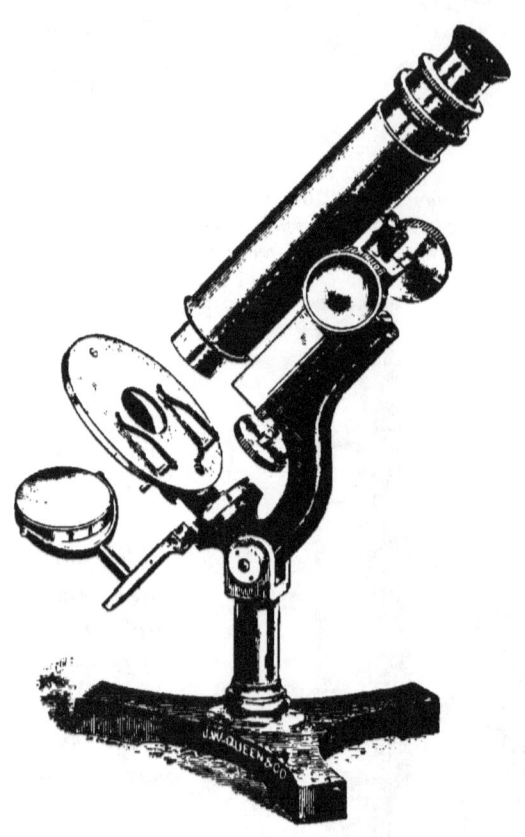

FIG. 84. *Queen & Company's Acme Microscope, No. 4.*

Fig. 85. *McIntosh Scientific Microscope, No. 2.*

CHAPTER III.

INTERPRETATION OF APPEARANCES.

APPARATUS AND MATERIAL FOR CHAPTER III.

A laboratory, compound microscope (§ 114); Preparation of fly's wing; 50 per cent. glycerin; Slides and covers; Preparation of letters in stairs (Fig. 86); Mucilage for air-bubbles and olive or clove oil for oil-globules (§ 127-130). Solid glass rod, and glass tube (§ 135-137); Collodion (§ 137); Carmine, India ink, or lamp black (§ 139-141); Frog, castor oil and micro-polariscope (§ 143).

INTERPRETATION OF APPEARANCES UNDER THE MICROSCOPE.

§ 120. **General Remarks.**—The experiments in this chapter are given secondarily for drill in manipulation, but primarily so that the student may not be led into error or puzzled by appearances which are constantly met with in microscopical investigation. Any one can look into a microscope, but it is quite another matter to interpret correctly the meaning of the appearances seen.

It is especially important to remember that the more of the relations of any object are known, the truer is the comprehension of the object. In microscopical investigation every object should be scrutinized from all sides and under all conditions in which it is likely to occur in nature and in microscopical investigation. It is best also to begin with objects of considerable size whose character is well known, to look at them carefully with the unaided eye so as to see them as wholes and in their natural setting. Then a low power is used, and so on step by step until the highest power available has been employed. One will in this way see less and less of the object as a whole, but every increase in magnification will give increased prominence to detail, detail which might be meaningless when taken alone and independent of the object as a whole. The pertinence of this advice will be appreciated when the student undertakes to solve the problems of histology; for even after all the years of incessant labor spent in trying to make out the structure of man and the lower animals, many details are still in doubt, the same visual appearances being quite differently interpreted by eminent observers.

Appearances which seem perfectly unmistakable with a low power may be found erroneous or very inadequate, for details of structure that were indistinguishable with the low power may become perfectly evident with a higher power or a more perfect objective. Indeed the problems of microscopic structure appear to become ever more complex, for difficulties overcome by improvements in the microscope simply give place to new difficulties, which in some cases render the subject more obscure than it appeared to be with the less perfect appliances.

The need of the most careful observation and constant watchfulness lest the appearances may be deceptive are thus admirably stated by Dallinger (See Carpenter-Dallinger, pp. 368–369) : " The correctness of the conclusions which the microscopist will draw regarding the nature of any object from the visual appearances which it presents to him when examined in the various modes now specified will necessarily depend in a great degree upon his previous experience in microscopic observation and upon his knowledge of the class of bodies to which the particular specimen may belong. Not only are observations of *any* kind liable to certain fallacies arising out of the previous notions which the observer may entertain in regard to the constitution of the objects or the nature of the actions to which his attention is directed, but even the most practiced observer is apt to take no note of such phenomena as his mind is not prepared to appreciate. Errors and imperfections of this kind can only be corrected, it is obvious, by general advance in scientific knowledge ; but the history of them affords a useful warning against hasty conclusions drawn from a too cursory examination. If the history of almost any scientific investigation were fully made known it would generally appear that the stability and completeness of the conclusions finally arrived at had been only attained after many modifications, or even entire alterations, of doctrine. And it is therefore of such great importance as to be almost essential to the correctness of our conclusions that they should not be finally formed and announced until they have been tested in every conceivable mode. It is due to science that it should be burdened with as few false facts [artifacts] and false doctrines as possible. It is due to other truth-seekers that they should not be misled, to the great waste of their time and pains, by our errors. And it is due to ourselves that we should not commit our reputation to the chance of impairment by the premature formation and publication of conclusions which may be at once reversed by other observers better informed than ourselves, or may be proved fallacious at some future time, perhaps even by our own more extended and careful researches. *The suspension of the judgment whenever there seems room for doubt* is a lesson inculcated by all those philosophers who have gained the highest repute for practical wisdom ; and it is one which the microscopist cannot too soon learn or too constantly practice."

For these experiments no condenser is to be used except where specifically indicated.

§ 121. **Dust or Cloudiness on the Ocular.**—Employ the 16 mm. (⅔ in.) objective, low ocular, and fly's wing as object.

Unscrew the field-lens and put some particles of lint from dark cloth on its upper surface. Replace the field-lens and put the ocular in position (§ 44). Light the field well and focus sharply. The image will be clear, but part of the field will be obscured by the irregular outline of the particles of lint. Move the object to make sure this appearance is not due to it.

Grasp the ocular by the milled ring, just above the tube of the microscope, and rotate it. The irregular object will rotate with the ocular. Cloudiness or particles of dust on any part of the ocular may be detected in this way.

§ 122. **Dust or Cloudiness on the Objective.**—Employ the same ocular and objective as before and the fly's wing as object. Focus and light well, and observe carefully the appearance. Rub glycerin on one side of a slide near the end. Hold the clean side of this end close against the objective. The image will be obscured, and cannot be made clear by focusing. Then use a clean slide, and the image may be made clear by elevating the tube slightly. The obscurity produced in this way is like that caused by clouding the front-lens of the objective. Dust would make a dark patch on the image that would remain stationary while the object or ocular is moved.

If a small diaphragm is employed and it is close to the object, only the central part of the field will be illuminated, and around the small light circle will be seen a dark ring (Fig. 42). If the diaphragm is lowered or a sufficiently large one employed the entire field will be lighted.

§ 123. **Relative Position of Objects or parts of the same object.** The general rule is that objects highest up come into focus *last* in focusing up, *first* in focusing down.

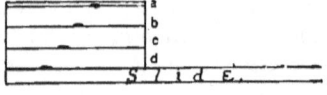

FIG. 86. *Letters mounted in stairs to show the order of coming into focus. a, b, c, d. The various letters indicated by the oblique row of black marks in the sectional view. Slide. The glass slide on which the letters are mounted.*

§ 124. **Objects having Plane or Irregular Outlines.**—As object use three printed letters in stairs mounted in Canada balsam (Fig. 86). The first letter is placed directly upon the slide, and covered with a small piece of glass about as thick as a slide. The second letter is placed upon this and covered in like manner. The third letter is placed upon the second thick cover and covered with an ordinary cover-glass. The letters should be as near together as possible, but not over-lapping. Employ the same ocular and objective as above (§ 121).

Lower the tube till the objective almost touches the top letter, then look into the microscope, and slowly focus up. The lowest letter will first appear, and then, as it disappears, the middle one will appear, and so on. Focus down, and the top letter will first appear, then the middle one, etc. The relative position of objects is determined exactly in this way in practical work.

For example, if one has a micrometer ruled on a cover-glass 15-25 hundredths mm. thick, it is not easy to determine with the naked eye which is the ruled surface. But if one puts the micrometer under a microscope and uses a 3 mm. ($\frac{1}{6}$th in.) objective, it is easily determined. The cover should be laid on a slide and focused till the lines are sharp. Now, without changing the focus in the least, turn the cover over. If it is necessary to focus up to get the lines of the micrometer sharp, the lines are on the upper side. If one must focus down, the lines are on the under surface. With a thin cover and delicate lines this method of determining the position of the rulings is of considerable practical importance.

§ 125. **Determination of the Form of Objects.**—The procedure is exactly as for the determination of the form of large objects. That is, one must examine the various aspects. For example, if one were placed in front of a wall of some kind he could not tell whether it was a simple wall or whether it was one side of a building unless in some way he could see more than the face of the wall. In other words, in order to get a correct notion of any body, one must examine more than one dimension,—two for plane surfaces, three for solids. So for microscopic objects, one must in some way examine more than one face. To do this with small bodies in a liquid the bodies may be made to roll over by pressing on one edge of the cover-glass. And in rolling over the various aspects are presented to the observer. With solid bodies, like the various organs, correct notions of the form of the elements can be determined by studying sections cut at right angles to each other. The methods of getting the elements to roll over, and of sectioning in different planes are in constant use in histology, and the microscopist who neglects to see all sides of the tissue elements has a very inadequate and often a very erroneous conception of their true form.

§ 126. **Transparent Objects having Curved Outlines.**—The success of these experiments will depend entirely upon the care and skill used in preparing the objects, in lighting, and in focusing.

Employ a 3 mm. (⅛ in.) or higher objective and a high ocular for all the experiments. It may be necessary to shade the object (§ 102) to get satisfactory results. When a diaphragm is used the opening should be small and it should be close to the object.

§ 127. **Air Bubbles.**—Prepare these by placing a drop of thin mucilage on the center of a slide and beating it with a scalpel blade until the mucilage looks milky from the inclusion of air bubbles. Put on a coverglass, but do not press it down.

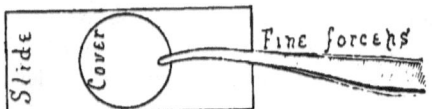

FIG. 87. *Diagram showing how to place a cover-glass upon an object with fine forceps.*

§ 128. **Air Bubbles with Central Illumination.**—Shade the object; and with the plane mirror, light the field with central light (Fig. 23).

Search the preparation until an air bubble is found appearing about 1 mm. in diameter, get it into the center of the field, and if the light is central the air bubble will appear with a wide, dark, circular margin and a small bright center. If the bright spot is not in the center, adjust the mirror until it is.

This is one of the simplest and surest methods of telling when the light is central or axial when no condenser is used (§ 61).

Focus both up and down, noting that, in focusing up, the central spot becomes very clear and the black ring very sharp. On elevating the tube of the microscope still more the center becomes dim, and the whole bubble loses its sharpness of outline.

§ 129. **Air Bubbles with Oblique Illumination.**—Remove the substage of the microscope and all the diaphragms. Swing the mirror so that the rays may be sent very obliquely upon the object (Fig. 23, C). The bright spot will appear no longer in the center but on the side *away from* the mirror (Fig. 88).

§ 130. **Oil Globules.**—Prepare these by beating a small drop of clove oil with mucilage on a slide and covering as directed for air bubbles (§ 128), or use a drop of milk.

§ 131. **Oil Globules with Central Illumination.**—Use the same diaphragm and light as above (§ 128). Find an oil globule appearing about 1 mm. in diameter. If the light is central the bright spot will appear in the center as with air. Focus up and down as with air, and note that the bright center of the oil globule is clearest *last* in focusing up.

FIG. 88. *Very small Globule of Oil (O) and an Air Bubble (A) seen by Oblique Light. The arrow indicates the direction of the light rays.*

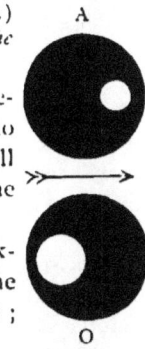

§ 132. **Oil Globules with Oblique Illumination.**—Remove the sub-stage, etc., as above, and swing the mirror to one side and light with oblique light. The bright spot will be eccentric, and will appear to be on the *same* side as the mirror (Fig. 88).

§ 133. **Oil and Air Together.**—Make a preparation exactly as described for air bubbles (§ 127), and add at one edge a little of the mixture of oil and mucilage (§ 130); cover and examine.

The sub-stage need not be used in this experiment. Search the preparation until an air bubble and an oil globule, each appearing about 1 mm. in diameter, are found in the same field of view. Light first with central light, and note that, in focusing up, the air bubble comes into focus first and that the central spot is smaller than that of the oil globule. Then, of course, the black ring will be wider in the air bubble than in the oil globule. Make the light oblique. The bright spot in the air bubble will move *away from* the mirror while that in the oil globule will move *toward it*. See Fig. 88.*

§ 134. **Air and Oil by Reflected Light.**—Cover the diaphragm or mirror so that no transmitted light (§ 60) can reach the preparation, using the same preparation as in § 133. The oil and air will appear like globules of silver on a dark ground. The part that was darkest in each will be lightest, and the bright central spot will be somewhat dark.†

§ 135. **Distinctness of Outline.**—In refraction images this depends on the difference between the refractive power of a body and that of the medium which surrounds it. The oil and air were very distinct in out-

---

* It should be remembered that the image in the compound microscope is inverted (Fig. 21), hence the bright spot really moves toward the mirror for air, and away from it for oil.

† It is possible to distinguish oil and air optically, as described above, only when quite high powers are used and very small bubbles are selected for observation. If a 16 mm. (⅔ in.) is used instead of a 3 mm. (⅛ in.) objective, the appearances will vary considerably from that given above for the higher power. It is well to use a low as well as a high power. Marked differences will also be seen in the appearances with objectives of small and of large aperture.

line as both differ greatly in refractive power from the medium which surrounds them, the oil being more refractive than the mucilage and the air less. (Figs. 53-55).

Place a fragment of a cover-glass on a clean slide, and cover it (see under mounting). The outline will be very distinct with the unaided eye. Use it as object and employ the 16 mm. (⅔ in.) objective and high ocular. Light with central light. The fragment will be outlined by a dark band. Put a drop of water at the edge of the cover-glass. It will run in and immerse the fragment. The outline will remain distinct, but the dark band will be somewhat narrower. Remove the cover-glass, wipe it dry, and wipe the fragment and slide dry also. Put a drop of 50% glycerin on the middle of the slide and mount the fragment of cover-glass in that. The dark contour will be much narrower than before.

Draw a solid glass rod out to a fine thread. Mount one piece in air, and the other in 50% glycerin. Put a cover-glass on each. Employ the same optical arrangement as before. Examine the one in air first. There will be seen a narrow, bright band, with a wide, dark band on each side.

The one in glycerin will show a much wider bright central band, with the dark borders correspondingly narrow (Fig. 89 b). The dark contour depends also on the numerical aperture of the objective—being wider with low apertures. This can be readily understood when it is remembered that the greater the aperture the more oblique the rays of light that can be received, and the dark band simply represents an area in which the rays are so greatly bent or refracted (Figs. 53, 55) that they cannot enter the objective and contribute to the formation of the image; the edges are dark simply because no light from them reaches the observer.

FIG. 89. *Solid glass rod showing the appearance when viewed with transmitted, central light, and with an objective of medium aperture.*
a. Mounted in air. b. Mounted in 50 per cent. glycerin.

If the glass rod or any other object were mounted in a medium of the same color and refractive power, it could not be distinguished from the medium.*

---

*Some of the rods have air bubbles in them, and then there results a capillary tube when they are drawn out. It is well to draw out a glass tube into a fine thread and examine it as described. The central cavity makes the experiment much more complex.

A very striking and satisfactory demonstration may be made by painting a zone or band of eosin or other transparent color on a solid glass rod, and immersing the rod in a test tube or vial of cedar oil, clove oil or turpentine. Above the liquid the glass rod is very evident, as it is also at the colored zone, but at other levels it can hardly be seen in the liquid.

§ 136. **Highly Refractive.**—This expression is often used in describing microscopic objects, (medulated nerve fibers, for example), and means that the object will appear to be bordered by a wide, dark margin when it is viewed by transmitted light. And from the above (§ 135), it would be known that the refractive power of the object, and the medium in which it was mounted must differ considerably.

§ 137. **Doubly Contoured.**—This means that the object is bounded by two, usually parallel dark lines with a lighter band between them. In other words, the object is bordered by (1) a dark line, (2) a light band, and (3) a second dark line (Fig. 90).

This may be demonstrated by coating a fine glass rod (§ 135) with one or more coats of collodion or celloidin and allowing it to dry, and then mounting in 50% glycerin as above. Employ a 3 mm. (⅛ in.) or higher objective, light with transmitted light, and it will be seen that where the glycerin touches the collodion coating there is a dark line— next this is a light band, and finally there is a second dark line where the collodion is in contact with the glass rod.* (Fig. 90).

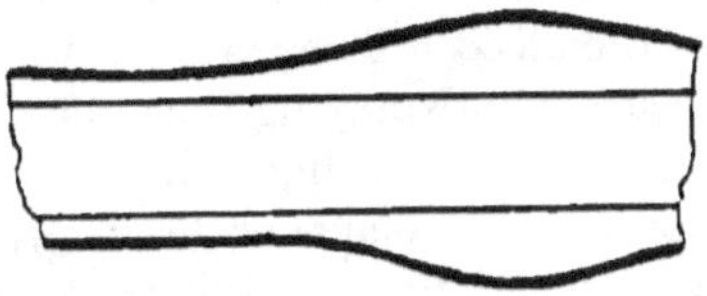

FIG. 90. *Solid glass rod coated with collodion to show a double contour. Toward one end the collodion had gathered in a fusiform drop.*

§ 138. **Optical Section.**—This is the appearance obtained in examining transparent or nearly transparent objects with a microscope when some plane below the upper surface of the object is in focus. The upper part of the object which is out of focus obscures the image but slightly. By changing the position of the objective or object, a different plane will be in focus and a different optical section obtained. The most satisfactory optical sections are obtained with high objectives having large aperture.

Nearly all the transparent objects studied may be viewed in optical

---

* The collodion used is a 6% solution of gun cotton in equal parts of sulphuric ether and 95% alcohol. It is well to dip the rod two or three times in the collodion and to hold it vertically while drying. The collodion will gather in drops, and one will see the difference between a thick and a thin membranous covering. (Fig. 90).

section. A striking example will be found in studying mammalian red blood-corpuscles on edge. The experiments with the solid glass rods (Fig. 89) furnish excellent and striking examples of optical sections.

§ 139. **Currents in Liquids.**—Employ the 16 mm. ($⅔$ in.) objective, and as object put a few particles of carmine on the middle of a slide, and add a drop of water. Grind the carmine well with a scalpel blade, and then cover it. If the microscope is inclined, a current will be produced in the water, and the particles of carmine will be carried along by it. Note that the particles seem to flow up instead of down—why is this?

Lamp-black rubbed in water containing a little mucilage answers well for this experiment.

§ 140. **Velocity Under the Microscope.**—In studying currents or the movement of living things under the microscope, one should not forget that the apparent velocity is as unlike the real velocity as the apparent size is unlike the real size. If one consults Fig. 37 it will be seen that the actual size of the field of the microscope with the different objectives and oculars is inversely as the magnification. That is, with great magnification only a small area can be seen. The field appears to be large, however, and if any object moves across the field it may appear to move with great rapidity, whereas if one measures the actual distance passed and notes the time, it will be seen that the actual motion is quite slow. One should keep this in mind in studying the circulation of the blood. The truth of what has just been said can be easily demonstrated in studying the circulation in the gills of Necturus, or in the frog's foot, by using first a low power in which the field is actually of considerable diameter (Fig. 37, Table, § 47) and then using a high power. With the high power the apparent motion will appear much more rapid. For the form of motion, spiral, serpentine, etc., see Carpenter-Dallinger, p. 375.

§ 141. **Pedesis or Brownian Movement.**—Employ the same object as above, but a 3 mm. ($⅛$ in.) or higher objective in place of the 16 mm. Make the body of the microscope vertical, so that there may be no currents produced. Use a small diaphragm and light the field well. Focus, and there will be seen in the field large motionless masses, and between them small masses in constant motion. This is an indefinite, dancing or oscillating motion.

This indefinite but continuous motion of small particles in a liquid is called *Pĕ-dĕ'sis or Brownian movement*. Also, but improperly, molecular movement, from the smallness of the particles.

The motion is increased by adding a little gum arabic solution or a slight amount of silicate of soda or of soap; sulphuric acid and various saline compounds retard or check the motion. One of the best objects is lamp-black ground up with a little gum arabic. Carmine prepared in the same way, or simply in water, is excellent; and very finely powdered pumice-stone in water has for many years been a favorite object.

Pedesis is exhibited by all solid matter if it is finely enough divided and in a suitable liquid. In the minds of most, no adequate explanation has yet been offered. See Carpenter-Dallinger, p. 373; Beale, p. 195; Jevons, in *Quart. Jour. Science*, n. s., Vol. VIII (1878), p. 167. In 1894 Meade Bache published a paper in the Proc. Amer. Philos. Soc., Vol. XXXIII, pp. 163–167, entitled "The Secret of the Brownian Movement." This paper is suggestive if not wholly satisfactory.

For the original account of this see Robert Brown, " Botanical appendix to Captain King's voyage to Australia," Vol. II, p. 534. (1826).

See also Dr. C. Aug. Sigm. Schultze, "Mikroskopische Untersuchungen über des Herren Robert Brown Entdeckung lebender, selbst im Feuer unzerstörbarer Theilchen in allen Körpern." From "Die Gesellschaft für Beförderung der Naturwissenschaften zu Freiburg." 1828.

Compare the pedetic motion with that of a current by slightly inclining the tube of the microscope. The small particles will continue their independent leaping movements while they are carried along by the current.

§ 142. **Demonstration of Pedesis with the Polarizing Microscope.**—The following demonstration shows conclusively that the pedetic motion is real and not illusive. (Ranvier, p. 173).

Open the abdomen of a dead frog (an alcoholic specimen will do if it is soaked in water for some time, but a fresh specimen is more satisfactory). Turn the viscera to one side and observe the small, whitish masses at the emergence of the spinal nerves. With fine forceps remove one of these and place it on the middle of a clean slide. Add a drop of water, or of water containing a little gum arabic. Rub the white mass around in the drop of liquid and soon the liquid will have a milky appearance. Remove the white mass, place a cover-glass on the milky liquid and seal the cover by painting a ring of castor oil all around it, half the ring being on the slide and half on the cover-glass. This is to avoid the production of currents by evaporation.

Put the preparation under the microscope and examine with, first a low then a high power (3 mm. or $\frac{1}{8}$ in.). In the field will be seen multitudes of crystals of carbonate of lime; the larger crystals are motionless but the smallest ones exhibit marked pedetic movement.

Use the micro-polariscope, light with great care and exclude all ad ventitious light from the microscope by shading the object (§ 102) and also by shading the eye. Focus sharply and observe the pedetic motion of the small particles, then cross the polarizer and analyzer, that is, turn one or the other until the field is dark. Part of the large, motionless crystals will shine continuously and a part will remain dark, but the small crystals between the large ones will shine for an instant, then disappear, only to appear again the next instant. This demonstration is believed to furnish absolute proof that the pedetic movement is real and not illusory.

§ 143. **Muscae Volitantes.**—These specks or filaments in the eyes due to minute shreds or opacities of the vitreous sometimes appear as part of the object as they are projected into the field of vision. They may be seen by looking into the well lighted microscope when there is no object under the microscope. They may also be seen by looking at the brightly illuminated snow or other white surface. By studying them carefully it will be seen that they are somewhat movable and float across the field of vision, and thus do not remain in one position as do the objects under observation. Furthermore, one may, by taking a little pains, familiarize himself with the special forms in his own eyes so that the more conspicuous, at least, may be instantly recognized.

§ 144. In addition to the above experiments it is very strongly recommended that the student follow the advice of Beale, p. 248, and examine first with a low then a higher power, mounted dry, then in water, lighted with reflected light, then with transmitted light, the following: Potato, wheat, rice, and corn starch, easily obtained by scraping the potato and the grains mentioned; bread crumbs; portions of feather. Portions of feather accidentally present in histological preparations have been mistaken for lymphatic vessels (Beale, 288). Fibers of cotton, linen and silk. Textile fibres accidentally present have been considered nerve fibres, etc. Human and animal hairs. Study with especial care hairs from various parts of the body of the animals used for dissection in the laboratory where you work. These are liable to be present in histological preparations, and unless their character is understood there is chance for much confusion and erroneous interpretation. The scales of butterflies and moths, especially the common clothes moth. The dust swept from carpeted and wood floors. Tea leaves and coffee grounds. Dust found in living rooms and places not frequently dusted. In the last will be found a regular museum of objects.

For figures (photo-micrographs, etc.) of the various forms of starch, see Bulletin No. 13 of the Chemical Division of the U. S. Department of Agriculture. For Hair and Wool, see Bulletin of the National Association of Wool Growers, 1875, p. 470, Proc. Amer. Micr. Soc., 1884, pp. 65-68.

For different appearances due to the illuminator, see Nelson, in Jour. Roy. Micr. Soc., 1891, pp. 90-105; and for the illusory appearances due to diffraction phenomena, see Carpenter-Dallinger, p. 376.

If it is necessary to see all sides of an ordinary gross object, and to observe it with varying illumination and under various conditions of temperature, moisture, etc., in order to obtain a fairly accurate and satisfactory knowledge of it, so much the more is it necessary not to be satisfied in microscopical observation until every means of investigation and verification has been called into service, and then of the image that falls upon the retina, only such details will be noted as the brain behind the eye is ready to appreciate.

To summarize this chapter and leave with the beginning student the result of the experience of many eminent workers:

1. Get all the information possible with the unaided eye. See the whole object and all sides of it, so far as possible.

2. Examine the preparation with a simple microscope in the same thorough way for additional detail.

3. Use a low power of the compound microscope.

4. Use a higher power.

5. Use the highest power available and applicable. In this way one sees the object as a whole and progressively more and more details. Then as the object is viewed from two or more aspects, something like a correct notion may be gained of its form and structure.

CHAPTER IV.

MAGNIFICATION AND MICROMETRY.

APPARATUS AND MATERIAL FOR THIS CHAPTER.

Simple and compound microscope; Steel scale or rule divided to millimeters and ½ths; Block for magnifier and compound microscope (§ 147, 151); Dividers (§ 147, 151); Stage micrometer (§ 150); Wollaston camera lucida (§ 151); Ocular screw-micrometer (Fig. 100); Micrometer ocular (Figs. 98–99). Abbe camera lucida (Fig. 96).

§ 145. **The Magnification, Amplification or Magnifying Power** of a simple or compound microscope is the ratio between the real and the apparent size of the object examined. The apparent size is obtained by measuring the virtual image (Figs. 21, 38). The object for determining magnification must be of known length and is designated *a micrometer* (§ 150). In practice a virtual image is measured by the aid of some form of camera lucida (Figs. 92, 96), or by double vision (§ 147). As the length of the object is known, the magnification is easily determined by dividing the apparent size of the image by the actual size of the object. For example, if the virtual image measures 40 mm. and the object magnified, 2 mm., the amplification must be $40 \div 2 = 20$, that is, the apparent size is 20 fold greater than the real size.

Magnification is expressed in diameters or times linear, that is, but one dimension is considered. In giving the scale at which a microscopical or histological drawing is made, the word magnification is frequently indicated by the sign of multiplication thus: × 450, upon a drawing would mean that the figure or drawing is 450 times as large as the object.

§ 146. **Magnification of Real Images.**—In this case the magnification is the ratio between the size of the real image and the size of the object, and the size of the real image can be measured directly. By recalling the work on the function of an objective (§ 49), it will be remembered that it forms a real image on the ground glass placed on the top of the tube, and that this real image could be looked at with the eye or measured as if it were an actual object. For example, suppose the object were 3 millimeters long and its image on the ground glass measured 15 mm., then the magnification must be, $15 \div 3 = 5$, that is, the real image is 5 times as long as the object. The real images seen in photography are mostly smaller than the objects, but the magnification is designated in the same way by dividing the size of the real image measured on the ground glass by the size of the object. For example, if the object is 400 millimeters long and its image on the ground glass is 25 mm. long, the ratio is $25 \div 400 = \frac{1}{16}$. That is, the image is $\frac{1}{16}$th as long as the object and is not magnified but reduced. In marking negatives, as with drawings, the sign of multiplication is put before the ratio, and in the example the designation would be $\times \frac{1}{16}$th.

MAGNIFICATION OF A SIMPLE MICROSCOPE.

§ 147. **The Magnification of a Simple Microscope** is the ratio between the object magnified (Fig. 16, A'B'), and the virtual image (A^3B^3). To obtain the size of this virtual image place the tripod magnifier near the edge of a support of such a height that the distance from the upper surface of the magnifier to the table is 250 millimeters.

As object, place a scale of some kind ruled in millimeters on the support under the magnifier. Put some white paper on the table at the base of the support, and on the side facing the light.

Close one eye, and hold the head so that the other will be near the upper surface of the lens. Focus if necessary to make the image clear (§ 9). Open the closed eye, and the image of the rule will appear as if on the paper at the base of the support. Hold the head very still, and, with dividers, get the distance between any two lines of the image. This is the so-called method of binocular or double vision in which the microscopic image is seen with one eye and the dividers with the other, the two images appearing to be fused in a single visual field.

§ 148. **Measuring the Spread of Dividers.**—This should be done on a steel scale divided to millimeters and ½ths.

As ½ mm. cannot be seen plainly by the unaided eye, place one arm of the dividers at a centimeter line, and then with the tripod magnifier count the number of spaces on the rule included between the points of the dividers. The magnifier simply makes it easy to count the spaces on the rule included between the points of the dividers—it does not, of course, increase the number of spaces or change their value.

As the distance between any two lines of the image of the scale gives the size of the virtual image (Fig. 16, $A^3 B^3$), and as the size of the object is known, the magnification is determined by dividing the size of image by the size of the object. Thus, suppose the distance between the two lines of the image is measured by the dividers and found on the steel scale to be 15 millimeters, and the actual size of the space between the two lines of the object is 2 millimeters, then the magnification must be $15 \div 2 = 7\frac{1}{2}$. That is, the image is $7\frac{1}{2}$ times as long or wide as the object. In this case the image is said to be magnified $7\frac{1}{2}$ diameters, or $7\frac{1}{2}$ times linear.

The magnification of any simple magnifier may be determined experimentally in the way described for the tripod.

MAGNIFICATION OF A COMPOUND MICROSCOPE.

§ 149. **The Magnification of a Compound Microscope** is the ratio between the final or virtual image (Fig. 21, $B^a A^a$), and the object magnified (A B).

The determination of the magnification of a compound microscope may be made as with a simple microscope (§ 147), but this is very fatiguing and unsatisfactory.

§ 150. **Stage, Object or Objective Micrometer.**—For determining the magnification of a compound microscope and for the purposes of micrometry, it is necessary to have a finely divided scale or rule on glass or on metal. Such a finely divided scale is called a micrometer, and for ordinary work one on glass is most convenient. The spaces between the lines should be $\frac{1}{10}$ and $\frac{1}{100}$ millimeter, and when high powers are to be used the lines should be very fine. It is of advantage to have the coarser lines filled with graphite (plumbago), especially when low powers are to be used. If one has an uncovered micrometer the lines may be very readily filled by rubbing some of the plumbago on the surface with the end of a cork; the superfluous plumbago may be removed by using a clean dry cloth or a piece of lens paper. After the lines are filled and the plumbago wiped from the surface, the slide should be examined and if it is found satisfactory, *i. e.*, if the lines are black, a cover-glass on which is a drop of warm balsam may be put over the lines to protect them.

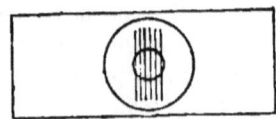

FIG. 91. *Diagram of a stage micrometer, with a ring on the lines to facilitate finding them.*

If one desires to have a part of the micrometer uncovered and a part covered for using homogeneous objectives, the lines may be filled with fine graphite, as described, and a piece of oblong cover-glass placed over a part of the band of lines.

§ 151. **Determination of Magnification.**—This is most readily accomplished by the use of some form of camera lucida (Ch. V), that of

Wollaston being most convenient as it may be used for all powers, and the determination of the *standard distance of 250 millimeters* at which to measure the image is very readily determined (Fig. 92, § 153).

Employ the 16 mm. (⅔ in.) objective and a 50 mm. (2 in., A. or No. 1) ocular and stage micrometer as object. For this power the $\tfrac{1}{10}$th mm. spaces of the micrometer should be used as object. Focus sharply, and make the tube of the microscope horizontal, by bending the flexible

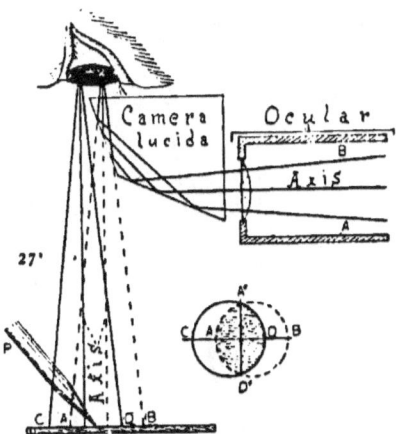

FIG. 92. *Wollaston's Camera Lucida, showing the rays from the microscope and from the drawing surface, and the position of the pupil of the eye.*
Axis, Axis. Axial rays from the microscope and from the drawing surface (Ch. V).
Camera Lucida. A section of the quadrangular prism showing the course of the rays in the prism from the microscope to the eye. As the rays are twice reflected, they have the same relation on entering the eye that they would have by looking directly into the ocular.
A B. The lateral rays from the microscope and their projection on the drawing surface.
C D. Rays from the drawing surface to the eye.
A D, A' D'. Overlapping portion of the two fields, where both the microscopic image and the drawing surface, pencil, etc., may be seen. It is represented by the shaded part in the overlapping circles at the right.
Ocular. The ocular of the microscope.
P. The drawing pencil. Its point is shown in the overlapping fields.

FIG. 92

pillar, being careful not to bring any strain upon the fine adjustment. (Frontispiece).

Put a Wollaston camera lucida (Ch. V.) in position, and turn the ocular around if necessary so that the broad flat surface may face directly upward, as shown in Fig. 92. Elevate the microscope by putting a block under the base, so that the perpendicular distance from the upper surface of the camera lucida to the table is 250 mm. (§ 153). Place some white paper on the work-table beneath the camera lucida.

Close one eye, and hold the head so that the other may be very close to the camera lucida. Look directly down. The image will appear to be on the table. It may be necessary to readjust the focus after the camera lucida is in position. If there is difficulty in seeing dividers and image consult Ch. V. Measure the image with dividers and obtain the power exactly as above (§ 147, 148).

Thus: Suppose two of the $\frac{1}{10}$th mm. spaces were taken as object, and the image is measured by the dividers, and the spread of the dividers is found on the steel rule to be $9\frac{2}{3}$ millimeters. If now the object is $\frac{2}{10}$ths of a millimeter and the magnified image is $9\frac{2}{3}$ millimeters, the magnification (which is the ratio between size of object and image) must be $9\frac{2}{3} \div \frac{2}{10} = 47$. That is, the magnification is 47 diameters, or 47 times linear. If the fractional numbers in the above example trouble the stu-

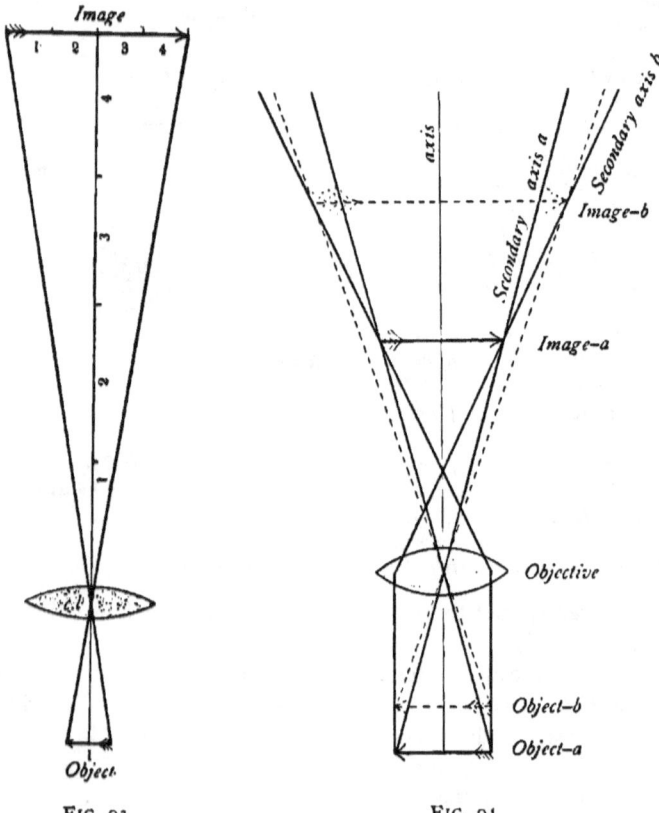

FIG. 93.   FIG. 94.

FIGS. 93-94. *Figures showing that the size of object and image vary directly as their distance from the center of the lens. In Fig. 94 one can also see why it is necessary to focus down, i. e., bring the object and objective nearer together when the tube is lengthened.*

dent, both may be reduced to the same denomination, thus: If the size of the image is found to be $9\frac{2}{5}$ mm. this number may be reduced to tenths mm., so it will be of the same denomination as the object. In 9 mm. there are 90 tenths, and in $\frac{2}{5}$ there are 4 tenths, then the whole length of the image is $90 + 4 = 94$ tenths of a millimeter. The object is 2 tenths of a millimeter, then there must have been a magnification of $94 \div 2 = 47$ diameters in order to produce an image 94 tenths of a millimeter long.

Put the 25 mm. (1 in. C, or No. 4) ocular in place of one of 50 mm. focus, and then put the camera lucida in position. Measure the size of the image with dividers and a rule as before. The power will be considerably greater than when the low ocular was used. This is because the virtual image (Fig. 21, $B^1 A^1$) seen with the high ocular is larger than the one seen with the low one. The real image (Fig. 21. $A^1 B^1$) remains nearly the same, and would be just the same if positive, par-focal oculars (§ 34, 68, note) were used.

Lengthen the tube of the microscope 50–60 mm. by pulling out the draw-tube. Remove the camera lucida, and focus, then replace the camera and obtain the magnification. It will be greater than with the shorter tube. This is because the real image (Fig. 94) is formed farther from the objective when the tube is lengthened, and the objective must be brought nearer the object (Fig. 94). The law is: The size of object and image varies directly as their distance from the center of the lens. The truth of this statement is illustrated by Figs. 93 and 94.

§ 152. **Varying the Magnification of a Compound Microscope.** It will be seen from the above experiments (§ 151) that independently of the distance at which the microscopic image is measured (§ 153), there are three ways of varying the power of a compound microscope. These are named below in the order of desirability.

(1) *By using a higher or lower objective.*
(2) *By using a higher or lower ocular.*
(3) *By lengthening or shortening the tube of the microscope* (Fig. 94).[*]

§ 153. **Standard Distance of 250 Millimeters at which the Virtual Image is Measured.**—For obtaining the magnification of both

---

[*] **Amplifier.**—In addition to the methods of varying the magnification given in § 152, the magnification is sometimes increased by the use of an amplifier, that is a diverging lens or combination placed between the objective and ocular and serving to give the image-forming rays from the objective an increased divergence. An effective form of this accessory was made by Tolles, who made it as a small achromatic concavo-convex lens to be screwed into the lower end of the draw-tube (frontispiece) and thus but a short distance above the objective. The divergence given to the rays increases the size of the real image about two-fold.

FIG. 95. *Figure showing the position of the microscope, the camera lucida, and the eye, and the different sizes of the image depending upon the distance at which it is projected from the eye. (a) The size at 25 cm.; (b) at 35 cm., (§ 153).*

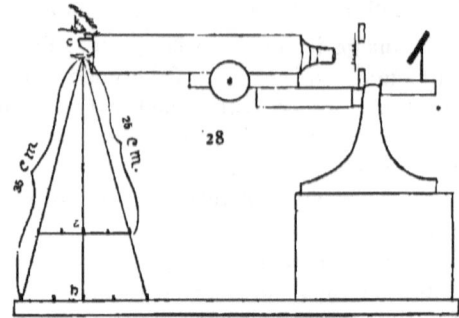

the simple and the compound microscope the directions were to measure the virtual image at a distance of 250 millimeters. This is not that the image could not be seen and measured at any other distance, but be-

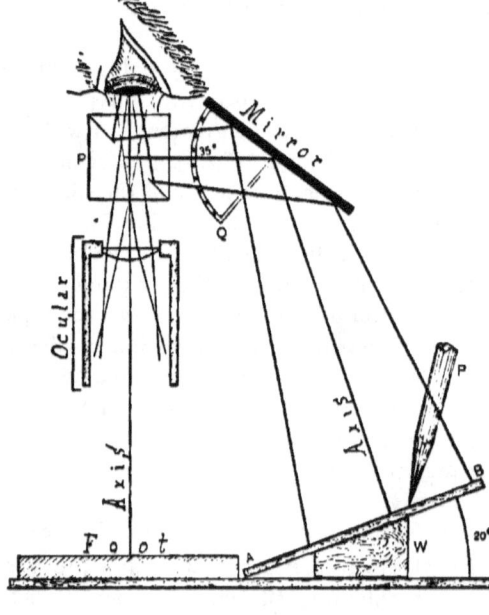

FIG. 96. *Sectional view of the Abbe Camera Lucida to show that in measuring the standard distance of 250 millimeters, one must measure along the axis from the point P, at the left of the prism, to the mirror, and from the mirror to the drawing surface. For a full explanation of this camera lucida, see next chapter, (Figs. 102, 106).*

FIG. 96.

cause some standard must be selected, and this is the most common one. The necessity for the adoption of some common standard will be seen at a glance in Fig. 95, where is represented graphically the fact that the size of the virtual image depends directly on the distance at which it is projected, and this size is directly proportional to the vertical distance from the apex of the triangle, of which it forms a base.

The distance of 250 millimeters has been chosen on the supposition that it is the distance of most distinct vision for the normal human eye.

Demonstrate the difference in magnification due to the distance at which the image is projected, by raising the microscope so that the distance will be 350 millimeters, then lowering to 150 millimeters.

In preparing drawings it is often of great convenience to make them at a distance somewhat less or somewhat greater than the standard. In such a case the magnification must be determined for the special distance. (See the next chapter.)

For discussions of the magnification of the microscope, see: Beale, pp. 41, 355; Carpenter-Dallinger, pp. 26, 238; Nägeli and Schwendener, p. 176, ; Ranvier, p. 29; Robin, p. 126; Amer. Soc. Micrs., 1884, p. 183; 1889, p. 22; Amer. Jour. Arts and Sciences, 1890, p. 50; Jour. Roy. Micr. Soc., 1888, 1889.

§ 154. **Table of Magnifications and of the Valuations of the Ocular Micrometer.**—*The following table should be filled out by each student. In using it for Micrometry and Drawing it is necessary to keep clearly in mind the exact conditions under which the determinations were made, and also the ways in which variation in magnification and the valuation of the ocular micrometer may be produced* (§ 152, 153, 163, 166).

OBJECTIVE.	OCULAR 50 mm.		OCULAR 25 mm.		OCULAR MICROMETER VALUATION.	
	TUBE IN	TUBE OUT —MM.	TUBE IN	TUBE OUT —MM.	TUBE IN.	OUT——MM.
	×	×	×	×		
	×	×	×	×		
	×	×	×	×		
	×	×	×	×		
	×	×	×	×		
	×	×	×	×		
SIMPLE MICROSCOPE.	×					

MICROMETRY.

§ 155. **Micrometry** is the determination of the size of objects by the aid of a microscope.

MICROMETRY WITH THE SIMPLE MICROSCOPE.

§ 156. With a simple microscope (A), the easiest and best way is to use dividers and then the simple microscope to see when the points of the dividers exactly include the object. The spread of the dividers is then obtained as above (§ 148). This amount will be the actual size of the object, as the microscope was only used in helping to see when the divider points exactly enclosed the object, and then for reading the divisions on the rule in getting the spread of the dividers.

(B) One may put the object under the simple microscope and then, as in determining the power (§ 147), measure the image at the standard distance. If now the size of the image so measured is divided by the magnification of the simple microscope, the quotient will give the actual size of the object.

Use a fly's wing, or some other object of about that size, and try to determine the width in the two ways described above. If all the work is accurately done the results will agree.

MICROMETRY WITH THE COMPOUND MICROSCOPE.

There are several ways of varying excellence for obtaining the size of objects with the compound microscope, the method with the ocular micrometer (§ 166–167) being most accurate.

§ 157. **Unit of Measure in Micrometry.**—As most of the objects measured with the compound microscope are smaller than any of the originally named divisions of the meter, and the common or decimal fractions necessary to express the size are liable to be unnecessarily cumbersome, *Harting*, in his work on the microscope (1859), proposed the **one thousandth of a millimeter** ($\frac{1}{1000}$ mm. or 0.001 mm.) or one millionth of a meter ($\frac{1}{1000000}$ or 0.000001 meter) as the unit. He named this unit micro-millimeter and designated it mmm. In 1869, *Listing* (Carl's Repetorium für Experimental-Physik, Bd, X, p. 5) favored the thousandth of a millimeter as unit and introduced the name **Mikron** or *micrum*. In English it is most often written *Micron*, plural *micra* or *microns*, pronunciation Mĭc'rŏn or Mīc'rŏn. By universal consent the sign or abbreviation used to designate it is the Greek μ. Adopting this unit and sign, one would express five thousandths of a millimeter ($\frac{5}{1000}$ or 0.005ths mm.) thus, 5μ.*

§ 158. **Micrometry** *by the use of a stage micrometer on which to mount the object.*—In this method the object is mounted on a micrometer and then put under the microscope, and the number of spaces covered by the object is read off directly. It is exactly like putting any large object on a rule and seeing how many spaces of the rule it covers. The defect in the method is that it is impossible to properly arrange objects on the micrometer. Unless the objects are circular in outline they are liable to be oblique in position, and in every case the end or edges of the object may be in the middle of a space instead of against one of the lines, consequently the size must be estimated or guessed at rather than really measured.

§ 159. **Micrometry** *by dividing the size of the image by the magnification of the microscope.*—For example, employ the 3 mm. (⅛ in.) objective, 25 mm. (1 in.) ocular, and a Necturus' red blood-corpuscle preparation as object. Obtain the size of the image of the long and short axes of three corpuscles with the camera lucida and dividers, exactly as in obtaining the magnification of the microscope (§ 151). Divide the size of the image in each case by the magnification, and the result will be the actual size of the blood-corpuscles. Thus, suppose the image of the long axis of the corpuscle is 18 mm. and the magnification of the microscope 400 diameters (§ 145), then the actual length of this long axis of the corpuscle is 18 mm. ÷ 400 = .045 mm. or 45 μ (§ 157).

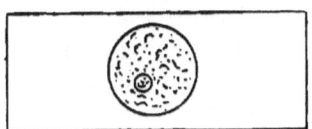

FIG. 97. *Preparation of blood with a ring around a group of blood corpuscles.*

As the same three blood-corpuscles are to be measured in three ways, it is an advantage to put a delicate ring around a group of three or more corpuscles, and make a sketch of the whole enclosed group, marking on the sketch the corpuscles measured. The different corpuscles vary considerably in size, so that accurate comparison of different methods of measurement can only be made when the same corpuscles are measured in each of the ways (Figs. 61–66).

*The term Micromillimeter ab. mmm. is very cumbersome, and besides is entirely inappropriate since the adoption of definite meanings for the prefixes *micro* and *mega*, meaning respectively one millionth and one million times the unit before which it is placed. A micromillimeter would then mean one-millionth of a millimeter, not one thousandth. The term micron has been adopted by the great microscopical societies, the international commission on weights and measures, and by original investigators, and is, in the opinion of the writer, the best term to employ. Jour. Roy. Micr. Soc., 1888, p. 502; Nature, Vol. XXXVII (1888), p. 388.

§ 160. **Micrometry** *by the use of a Stage Micrometer and a Camera Lucida.*—Employ the same object, objective and ocular as before. Put the camera lucida in position, and with a lead pencil make dots on the paper at the limits of the image of the blood-corpuscle. Measure the same three that were measured in § 159.

Remove the object, place the stage micrometer under the microscope, focus well, and draw the lines of the stage micrometer so as to include the dots representing the limits of the part of the image to be measured. As the value of the spaces on the stage micrometer is known, the size of the object is determined by the number of spaces of the micrometer required to include it.

This simply enables one to put the image of a fine rule on the image of a microscopic object. It is theoretically an excellent method, and nearly the same as measuring the spread of the dividers with a simple microscope (§§ 148, 167).

OCULAR MICROMETER.

§ 161. **Ocular Micrometer, Eye-Piece Micrometer.**—This, as the name implies, is a micrometer to be used with the ocular. It is a micrometer on glass, and the lines are sufficiently coarse to be clearly seen by the ocular. The lines should be equidistant and about $\frac{1}{10}$th or $\frac{1}{20}$th mm. apart, and every fifth line should be longer and heavier to facilitate counting. If the micrometer is ruled in squares (*net-micrometer*) it will be very convenient for many purposes.

The ocular micrometer is placed in the ocular, no matter what the form of the ocular (*i. e.*, whether positive or negative) at the level at which the real image is formed by the objective, and the image appears to be immediately upon or under the ocular micrometer, and hence the number of spaces on the ocular micrometer required to measure the real image may be read off directly. This is measuring the size of the real image, however, and the actual size of the object can only be determined by determining the ratio between the size of the real image and the object. In other words, it is necessary to get the *valuation of the ocular micrometer* in terms of a stage micrometer.

§ 162. **Valuation of the Ocular Micrometer.**—This is the value of the divisions of the ocular micrometer for the purposes of micrometry, and is entirely relative, depending upon the magnification of the real image formed by the objective, consequently it changes with every change in the magnification of the real image, and must be specially determined for every optical combination (*i. e.*, objective and ocular), and for every change in the length of the tube of the microscope. That is, it is necessary to determine the ocular micrometer valuation for every condition modifying the real image of the microscope (§ 152).

Any Huygenian ocular (Fig 30) may, however, be used as a micrometer ocular by placing the ocular micrometer at the level of the ocular diaphragm, where the real image is formed. If there is a slit in the side of the ocular, and the ocular micrometer is mounted in some way it may be introduced through the opening in the side. When no side opening exists the mounting of the eye-lens may be unscrewed and the ocular micrometer, if on a cover-glass, can be laid on the upper side of the ocular diaphragm.

FIGS. 98, 99. *Filar Ocular Micrometer with Field (Bausch & Lomb, Optical Co.). For other ocular micrometers see pp. 25, 26.*

§ 163. **Obtaining the Valuation of the Filar Micrometer.** This micrometer (Fig. 98-99) consists of a Ramsden's ocular and cross lines. As seen in Fig. 98 there are three lines. The horizontal and one vertical line are fixed. One vertical line may be moved by the screw back and forth across the field.

For obtaining the valuation of this ocular micrometer an accurate stage micrometer must be used. Carefully focus the $\frac{1}{10}$th mm. spaces. The lines of the ocular micrometer should also be sharp. If they are not focus them by moving the top of the ocular up or down (§ 164). Make the vertical lines of the filar micrometer parallel with the lines of the stage micrometer. Take the precautions regarding the width of the stage micrometer lines given in § 167 (see also Fig. 101). Note the position of the graduated wheel and of the teeth of the recording comb, and then rotate the wheel until the movable line traverses one space on the stage micrometer. Each tooth of the recording comb indicates a total revolution of the wheel, and by noting the number of teeth required and the graduations on the wheel, the revolutions and parts of revolutions required to measure the $\frac{1}{10}$ th mm. of the stage micrometer can be easily noted. Measure in like manner 4 or 5 spaces and get the average. Suppose this average is $1\frac{1}{4}$ th revolutions or 125 graduations on the wheel, to measure the $\frac{1}{100}$th mm. or 10μ (see § 157), then one of the graduations on the wheel would measure 10μ divided by 125 = $.08\mu$. In using this valuation for actual measurement, the tube of the microscope and the objective must be exactly as when obtaining the valuation (see § 165).

Example of Measurement.—Suppose one uses the red blood corpuscles of a dog or monkey, etc., every condition being as when the valuation was determined, one notes very accurately how many of the graduations on the wheel are required to make the movable line traverse the object from edge to edge. Suppose it requires 94 of the graduations to measure the diameter, the actual size of the corpuscle would be 94 × .08μ = 7.52μ.

The advantage of the filar micrometer is that the valuation of one graduation being so small, even the smallest object to be measured would require several graduations to measure it. In ocular micrometers with fixed lines, small objects like bacteria might not fill even one space, therefore estimations, not measurements, must be made. For large objects, like most of the tissue elements, the ocular micrometers with fixed lines answer very well, for the part which must be estimated is relatively small, and the chance of error is correspondingly small.

§ 164. **Obtaining the Ocular Micrometer Valuation for an Ocular Micrometer with Fixed Lines** (Figs. 33, 34, p. 25).—Use the stage micrometer as object. Light the field well and look into the microscope. The lines of the ocular micrometer should be very sharply defined. If they are not raise or lower the eye-lens to make them so; that is, focus as with the simple magnifier.

When the lines of the ocular micrometer are distinct, focus the microscope (§ 45, 46, 56) for the stage micrometer. The image of the stage micrometer will appear to be directly under or upon the ocular micrometer.

Make the lines of the two micrometers parallel by rotating the ocular or changing the position of the stage micrometer, or both if necessary, and then make any two lines of the stage micrometer coincide with any two on the ocular micrometer. To do this it may be necessary to pull out the draw-tube a greater or less distance. See how many spaces are included on each of the micrometers.

Divide the value of the included space or spaces on the stage micrometer by the number of divisions on the ocular micrometer required to include them, and the quotient so obtained will give the valuation of the ocular micrometer in fractions of the unit of measure of the stage micrometer. For example, suppose the millimeter is taken as the unit for the stage micrometer and this unit is divided into spaces of $\frac{1}{10}$th and $\frac{1}{100}$th millimeter. If now, with a given optical combination and tube-length, it requires 10 spaces on the ocular micrometer to include the real image of $\frac{1}{10}$th millimeter on the stage micrometer, obviously one space on the ocular micrometer would include only one-tenth as much, or $\frac{1}{10}$th mm. ÷ 10 = $\frac{1}{100}$th mm. That is, each space on the ocular micrometer would include $\frac{1}{100}$th of a millimeter on the stage micrometer, or $\frac{1}{100}$th millimeter of length of any object under the microscope, the conditions remaining the same. Or, in other words, it would require

100 spaces on the ocular micrometer to include 1 millimeter on the stage micrometer, then as before 1 space of the ocular micrometer would have a valuation of $\frac{1}{100}$th millimeter for the purposes of micrometry; and the size of any minute object may be determined by multiplying this valuation of one space by the number of spaces required to include it. For example, suppose the fly's wing or some part of it covered 8 spaces on the ocular micrometer, it would be known that the real size of the part measured is $\frac{1}{100}$th mm. $\times$ 8 = $\frac{8}{100}$ mm. or 80 μ (§ 157).

§ 165. **Varying the Ocular Micrometer Valuation.**—Any change in the objective, the ocular or the tube-length of the microscope, that is to say, any change in the size of the real image, produces a corresponding change in the ocular micrometer valuation (§ 152, 161).

§ 166. **Micrometry with the Ocular Micrometer.**—Use the 3 mm. ($\frac{1}{8}$ in.) objective and preparation of Necturus blood-corpuscles as object. Make certain that the tube of the microscope is of the same length as when determining the ocular micrometer valuation. In a word, be sure that all the conditions are exactly as when the valuation was determined, then put the preparation under the microscope and find the same three red corpuscles that were measured in the other ways (§ 159, 160).

Count the divisions on the ocular micrometer required to enclose or measure the long and the short axis of each of the three corpuscles, then multiply the number of spaces in each case by the valuation of the ocular micrometer for this objective, tube-length and ocular, and the results will represent the actual length of the axes of the corpuscles in each case.

The same corpuscle is, of course, of the same actual size, when measured in each of the three ways, so that if the methods are correct and the work carefully enough done, the same results should be obtained by each method. See general remarks on micrometry (§ 167).*

* There are three ways of using the ocular micrometer, or of arriving at the size of the objects measured with it :

(A) By finding the value of a division of the ocular micrometer for each optical combination and tube-length used, and employing this valuation as a multiplier. This is the method given in the text, and is the one most frequently employed. Thus, suppose with a given optical combination and tube-length it required five divisions on the ocular micrometer to include the image of $\frac{1}{10}$ths millimeter of the stage micrometer, then obviously one space on the ocular micrometer would include $\frac{1}{5}$th of $\frac{1}{10}$ths mm. or $\frac{1}{25}$th mm.; and the size of any unknown object under the microscope would be obtained by multiplying the number of divisions on the ocular micrometer required to include its image by the value of one space, or in this case, $\frac{1}{25}$th mm. Suppose some object, as the fly's wing, required 15 spaces of the ocular micrometer to include some part of it, then the actual size of this part of the wing would be 15 $\times$ $\frac{1}{25}$ = $\frac{3}{5}$ths, or 0.6 mm.

§ 167. Remarks on Micrometry.—In using adjustable objectives (§ 22, 96), the magnification of the objective varies with the position of the adjusting collar, being greater when the adjustment is closed as for thick cover glasses than when open, as for thin ones. This variation in the magnification of the objective produces a corresponding change in the magnification of the entire microscope and the ocular micrometer valuation—therefore it is necessary to determine the magnification and ocular micrometer valuation for each position of the adjusting collar.

While the principles of micrometry are simple, it is very difficult to get the exact size of microscopic objects. This is due to the lack of perfection and uni-

(B) By finding the number of divisions on the ocular micrometer required to include the image of an entire millimeter of the stage micrometer, and using this number as a divisor. This number is also sometimes called the *ocular micrometer ratio*. Taking the same case as in (A), suppose five divisions of the ocular micrometer are required to include the image of $\frac{2}{10}$ths mm., on the stage micrometer, then evidently it would require $5 \div \frac{2}{10} = 25$ divisions on the ocular micrometer to include a whole millimeter on the stage micrometer, then the number of divisions of the ocular micrometer required to measure an object divided by 25 would give the actual size of the object in millimeters or in a fraction of a millimeter. Thus, suppose it required 15 divisions of the ocular micrometer to include the image of some part of the fly's wing, the actual size of the part included would be $15 \div 25 = \frac{3}{5}$ or 0.6 mm. This method is really exactly like the one in (A), for dividing by 25 is the same as multiplying by $\frac{1}{25}$th.

(C) By having the ocular micrometer ruled in millimeters and divisions of a millimeter, and then getting the size of the real image in millimeters. In employing this method a stage micrometer is used as object and the size of the image of one or more divisions is measured by the ocular micrometer, thus: Suppose the stage micrometer is ruled in $\frac{1}{10}$th and $\frac{1}{100}$th mm. and the ocular micrometer as ruled in millimeters and $\frac{1}{10}$th mm. Taking $\frac{2}{10}$th mm. on the stage micrometer as object, as in the other cases, suppose it requires 10 of the $\frac{1}{10}$th mm. spaces or 1 mm. to measure the real image, then the real image must be magnified $\frac{10}{10} \div \frac{2}{10} = 5$ diameters, that is, the real image is five times as great in length as the object, and the size of an object may be determined by putting it under the microscope and getting the size of the real image in millimeters with the ocular micrometer and dividing it by the magnification of the real image, which in this case is 5 diameters.

Use the fly's wing as object, as in the other cases, and measure the image of the same part. Suppose that it required 30 of the $\frac{1}{10}$ mm. divisions $\frac{30}{10}$ mm or 3 mm. to include the image of the part measured, then evidently the actual size of the part measured would be 3 mm. $\div 5 = \frac{3}{5}$ mm., the same result as in the other cases.

In comparing these methods it will be seen that in the first two (A and B) the ocular micrometer may be simply ruled with equidistant lines without regard to the absolute size in millimeters or inches of the spaces. In the last method the ocular micrometer must have its spaces some known division of a millimeter or inch. In the first two methods only one standard of measure is required, viz., the stage micrometer; in the last method two standards must be used,—a stage micrometer and an ocular micrometer. Of course, the ocular micrometer in the first two cases must have the lines equidistant as well as in the last case, but ruling lines equidistant and an exact division of a millimeter or an inch are two quite different matters.

formity of micrometers, and the difficulty of determining the exact limits of the object to be measured. Hence, all microscopic measurements are only approximately correct, the error lessening with the increasing perfection of the apparatus and the skill of the observer.

A difficulty when one is using high powers is the width of the lines of the micrometer. If the micrometer is perfectly accurate half the width of each line belongs to the contiguous spaces, hence one should measure the image of the space from the centers of the lines bordering the space, or as this is somewhat difficult in using the ocular micrometer, one may measure from the inside of one bordering line and from the outside of the other. If the lines are of equal width this is as accurate as measuring from the center of the lines. Evidently it would not be right to measure from either the inside or the outside of both lines (Fig. 101).

It is also necessary in micrometry to use an objective of sufficient power to enable one to see all the details of an object with great distinctness. The necessity of using sufficient amplification in micrometry has been especially remarked upon by Richardson, Monthly Micr. Jour., 1874, 1875 ; Rogers, Proc. Amer. Soc. Microscopists, 1882, p. 239; Ewell, North American Pract., 1890, pp. 97, 173.

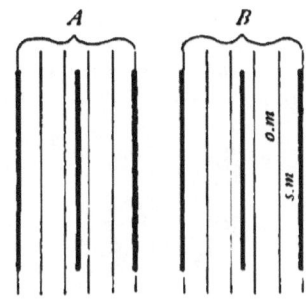

Fig. 101. *The appearance of the coarse stage and of the fine ocular micrometer lines when using a high objective.*

(*A*). *The method of measuring the spaces by putting the fine ocular micrometer lines opposite the center of the coarse stage micrometer lines.*

(*B*). *Method of measuring the spaces of the stage micrometer by putting one line of the ocular micrometer (o.m.) at the inside and one at the outside of the coarse stage micrometer lines (s.m.).*

FIG. 101.

As to the limit of accuracy in micrometry, one who has justly earned the right to speak with authority, expresses himself as follows : "*I assume that 0.2µ is the limit of precision in microscopic measures, beyond which it is impossible to go with certainty.*" W. A. Rogers, Proc. Amer. Soc Micrs., 1883, p. 198.

In comparing the methods of micrometry with the compound microscope, given above (158, 159, 160, 166), the one given in ¿ 158 is impracticable, that given in ¿ 159 is open to the objection that two standards are required,—the stage micrometer, and the steel rule ; it is open to the further objection that several different operations are necessary, each operation adding to the probability of error. Theoretically the method given in ¿ 160 is good, but it is open to the very serious objection in practice that it requires so many operations which are especially liable to introduce errors. The method that experience has found most safe and expeditious, and applicable to all objects, is the method with the ocular micrometer. If the valuation of the ocular micrometer has been accurately determined, then the only difficulty is in deciding on the exact limits of the object to be measured and so arranging the ocular micrometer that these limits are inclosed by some divisions of the micrometer. Where the object is not exactly included by whole spaces on the ocular micrometer, the chance of error comes in, in estimating just how far into a space the object reaches on the side not in contact with one of the micrometer lines. If the ocular micrometer has some quite narrow spaces, and others considerably larger, one can nearly always manage to exactly include the object by some two lines. The ocular screw-micrometer (Fig. 100) obviates this entirely as the cross hairs or lines traverse the object or its real image, and whether this distance be great or small it can be read off on the graduated wheel, and no estimation or guess work is necessary.

For those especially interested in micrometry, as in its relation to medical jurisprudence, the following references are recommended. These articles consider the problem in a scientific as well as a practical spirit: The papers of Prof. Wm. A. Rogers on micrometers and micrometry, in the Amer. Quar. Micr. Jour., Vol. I, pp. 97, 208; Proceedings Amer. Soc. Microscopists, 1882, 1883, 1887. Dr. M. D. Ewell, Proc. Amer. Soc. Micrs., 1890; The Microscope, 1889, pp. 43-45; North Amer. Pract., 1890, pp. 97, 173. Dr. J. J. Woodward, Amer. Jour. of the Med. Sci., 1875. M. C. White, Article Blood-stains, Ref. Hand-Book, Med. Sciences, 1885. Medico-Legal Journal, Vol. XII. For the change in magnification due to a change in the adjustment of adjustable objectives, see Jour. Roy. Micr. Soc., 1880, p. 702; Amer. Monthly Micr. Jour., 1880, p. 67.

If one consults the medico-legal journals, the Index Medicus, and the Index Catalog of the Library of the Surgeon General's Office, under Micrometry, Blood, and Jurisprudence, he can get on track of the main work which has been and is being done.

CHAPTER V.

DRAWING WITH THE MICROSCOPE.

APPARATUS AND MATERIAL FOR THIS CHAPTER.

Microscope, Abbe camera lucida, drawing board thumb tacks, pencils, paper, and microscope screen (Fig. 58).

DRAWING MICROSCOPIC OBJECTS.

§ 168. Microscopic objects may be drawn free-hand directly from the microscope, but in this way a picture giving only the general appearance and relations of parts is obtained. For pictures which shall have all the parts of the object in true proportions and relations, it is necessary to obtain an exact outline of the image of the object, and to locate in this outline all the principal details of structure. It is then possible to complete the picture free-hand from the appearance of the object under the microscope. The appliance used in obtaining outlines, etc., of the microscopic image is known as a *camera lucida*.

§ 169. **Camera Lucida.**—This is an optical apparatus for enabling one to see objects in greatly different situations, as if in one field of vision, and with the same eye. In other words, it is an optical device for superimposing or combining two fields of view in one eye.

As applied to the microscope, it causes the magnified virtual image of the object under the microscope to appear as if projected upon the table or drawing board, where it is visible with the drawing paper, pencils, dividers, etc., by the same eye, and in the same field of vision. The microscopic image appears like a picture on the drawing paper. This is accomplished in two distinct ways:

(A) By a camera lucida reflecting the rays from the microscope so that their direction when they reach the eye coincides with that of the rays from the drawing paper, pencils, etc. In some of the camera lucidas of this group (Wollaston's, Fig. 105), the rays are reflected twice, and the image appears as when looking directly into the microscope. In others the rays are reflected but once, and the image has the inversion produced by a plane mirror. For drawing purposes this inversion is a great objection, as it is necessary to similarly invert all the details added free-hand.

(B) By a camera lucida reflecting the rays of light from the drawing paper, etc., so that their direction when they reach the eye coincides with the direction of the rays from the microscope (Fig. 57, 60). In all of the camera lucidas of this group, the rays from the paper are twice reflected and no inversion appears.

The better forms of camera lucidas (Wollaston's, Grunow's, Abbe's, etc.), may be used for drawing both with low and with high powers. Some require the microscope to be inclined (Fig. 105), while others are

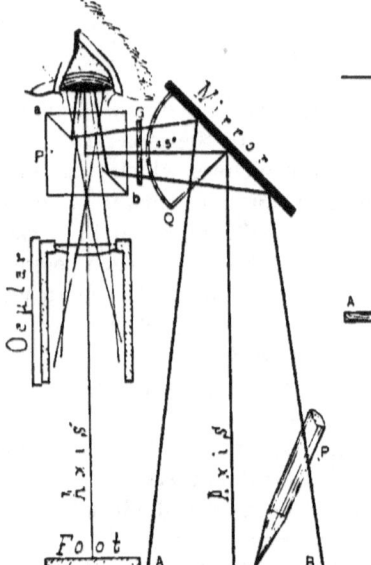

FIG. 102.

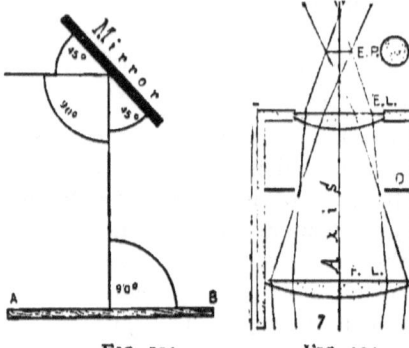

FIG. 103.   FIG. 104.

FIG. 102. *Abbe Camera Lucida with the mirror at 45°, the drawing surface horizontal, and the microscope vertical.*

Axis, Axis. *Axial ray from the microscope and from the drawing surface.*
A B. *Marginal rays of the field on the drawing surface.* a b. *Sectional view of the silvered surface on the upper of the triangular prisms composing the cubical prism (P). The silvered surface is shown as incomplete in the center, thus giving passage to the rays from the microscope.*
Foot. *Foot or base of the microscope.*
G. *Smoked glass seen in section. It is placed between the mirror and the prism to reduce the light from the drawing surface.*
Mirror. *The mirror of the camera lucida. A quadrant (Q) has been added to indicate the angle of inclination of the mirror, which in this case is 45°.*
Ocular. *Ocular of the microscope over which the prism of the camera lucida is placed.*
P, P. *Drawing pencil and the cubical prism over the ocular.*

FIG. 103. *Geometrical figure showing the angles made by the axial ray with the drawing surface and the mirror.*
A B. *The drawing surface.*

FIG. 104. *Ocular showing eye-point, E P. It is at this point both horizontally and vertically that the hole in the silvered surface should be placed (§ 173).*

designed to be used on the microscope in a vertical position. As in biological work, it is often necessary to have the microscope vertical, the form for a vertical microscope is to be preferred ; but see Figs. 102–111.

§ 170. **Avoidance of Distortion.**—*In order that the picture drawn by the aid of a camera lucida may not be distorted, it is necessary that the axial ray from the image on the drawing surface shall be at right angles to the drawing surface* (Figs. 102, 105, 106).

§ 171. **Wollaston's Camera Lucida.**—This is a quadrangular prism of glass put in the path of the rays from the microscope, and it serves to change the direction of the axial ray 90 degrees. In using it the microscope is made horizontal, and the rays from the microscope enter one half of the pupil while rays from the drawing surface enter the other half of the pupil. As seen in the figure (Fig. 105), the fields partly overlap, and where they do so overlap, pencil or dividers and microscopic image can be seen together.

In drawing or using the dividers with the Wollaston camera lucida it is necessary to have the field of the microscope and the drawing surface about equally lighted. If the drawing surface is too brilliantly lighted the pencil or dividers may be seen very clearly, but the microscopic image will be obscure. On the other hand, if the field of the microscope has too much light the microscopic image will be very definite, but the pencil or dividers will not be visible. It is necessary, as with the Abbe camera lucida (§ 173), to have the Wollaston prism properly arranged with reference to the axis of the microscope and the eye-point. If it is not, one will be unable to see the image well, and may be entirely unable to see the pencil and the image at the same time. Again, as rays from the microscope and from the drawing surface must enter independent parts of the pupil of the same eye, one must hold the eye so that the pupil is partly over the camera lucida and partly over the drawing surface. One can tell the proper position by trial. This is not a very satisfactory camera to draw with, but it is a very good form to measure the vertical distance of 250 mm. at which the drawing surface should be placed when determining magnification (§ 153).

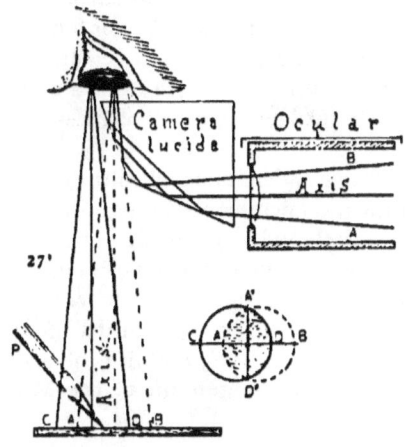

FIG. 105. *Wollaston's Camera Lucida, showing the rays from the microscope and from the drawing surface, and the position of the pupil of the eye.*

For full explanation see Fig. 92.

§ 172. **Abbe Camera Lucida.**—This consists of a cube of glass cut into two triangular prisms and silvered on the upper one. A small oval hole is then cut out of the center of the silvered surface and the two prisms are cemented together, thus giving a cubical prism with a perforated 45 degree mirror (Fig.

102, a b). The upper surface of the prism is covered by a perforated metal plate (Fig. 108). This prism is placed over the ocular in such a way that the light from the microscope passes through the hole in the silvered face and thence directly to the eye. Light from the drawing surface is reflected by a mirror to the silvered surface of the prism and reflected by this surface to the eye in company with the rays from the microscope, so that the two fields appear as one, and the image is seen as if on the drawing surface (Figs. 102, 106). It is designed for use with a vertical microscope, but see § 174.

§ 173. **Arrangement of the Camera Lucida Prism.**—In placing this camera lucida over the ocular for drawing or the determination of magnification, the center of the hole in the silvered surface is placed in the optic axis of the microscope. This is done by properly arranging the centering screws that clamp the camera to the microscope tube or ocular. The perforation in the silvered surface must also be at the level of the eye-point (Fig. 104). In other words, the prism must be so arranged vertically and horizontally that the hole in the silvered surface will be in the axis of the microscope and co-incident with the eye-point of the ocular. If it is above or below, or to one side of the eye-point, part or all of the field of the microscope will be cut off. As stated above, the centering screws are for the proper horizontal arrangement of the prism. The prism is set at the right height by the makers for the eye-point of a medium ocular. If one desires to use an ocular with the eye-point farther away or nearer, as in using high or low oculars, the position of the eye-point may be determined as directed in § 55 and the prism loosened and raised or lowered to the proper level; but in doing this one should avoid setting the prism obliquely to the mirror.

In the latest and best forms of this camera lucida special arrangements have been made for raising or lowering the prism so that it may be used with equal satisfaction on oculars with the eye-point at different levels, and the prism is hinged to turn aside without disturbing the mirror.

See the latest catalogs of Zeiss, Leitz, and the Bausch & Lomb Optical Co.

One can determine when the camera is in a proper position by looking into the microscope through it. If the field of the microscope appears as a circle and of about the same size as without the camera lucida, then the prism is in a proper position. If one side of the field is dark, then the prism is to one side of the center; if the field is considerably smaller than when the prism is turned off the ocular, it indicates that it is not at the correct level, i. e., it is above or below the eye-point.

§ 174. **Arrangement of the Mirror and the Drawing Surface.**—

The Abbe camera lucida was designed for use with a vertical microscope (Fig. 102). On a vertical microscope, if the mirror is set at an angle of 45°, the axial ray will be at right angles with the table top or a drawing board which is horizontal, and a drawing made under these conditions would be in true proportion and not distorted. The stage of most microscopes, however, extends out so far at the sides that with a 45° mirror the image appears in part on the stage of the microscope. In order to avoid this the mirror may be depressed to some point below 45°, say at 40° or 35° (Fig. 106-107). But as the axial ray from the mirror to the prism must still be reflected horizontally, it follows that the axial ray will no longer form an angle of 90 degrees with the drawing surface, but a greater angle. If the mirror is depressed to 35°, then the axial ray must take an angle of 110° with a horizontal drawing surface; see the geometrical figure, (Fig. 107). To make the angle 90° again, so that there shall be no distortion, the drawing board must be raised toward the microscope 20°. **The general rule is to raise the drawing board twice as many degrees toward the microscope as the mirror is depressed below 45°.** Practically the field for drawing can always be made free of the stage of the microscope, at 45°, at 40°, or at 35°. In the first case (45° mirror) the drawing surface should be horizontal, in the second case (40° mirror) the drawing surface should be elevated 10°, and in the third case (35° mirror) the drawing board should be elevated 20° toward the microscope. Furthermore it is necessary in using an elevated drawing board to have the mirror bar project directly laterally so that the edges of the mirror will be in planes parallel with the edges of the drawing board, otherwise there will be front to back distortion, although the elevation of the drawing board would avoid right to left distortion. If one has a micrometer ruled in squares (*net micrometer*) the distortion produced by not having the axial ray at right angles with the drawing surface may be very strikingly shown. For example, set the mirror at 35° and use a horizontal drawing board. With a pencil make dots at the corners of some of the squares, and then with a straight edge connect the dots. The figures will be considerably longer from right to left than from front to back. Circles in the object would appear as ellipses in the drawings, the major axis being from right to left.

The angle of the mirror may be determined with a protractor, but that is troublesome. It is much more satisfactory to have a quadrant attached to the mirror and an indicator on the projecting arm of the mirror. If the quadrant is graduated throughout its entire extent, or preferably at three points, 45°, 40° and 35°, one can set the mirror at a

FIG. 108.   FIG. 107.

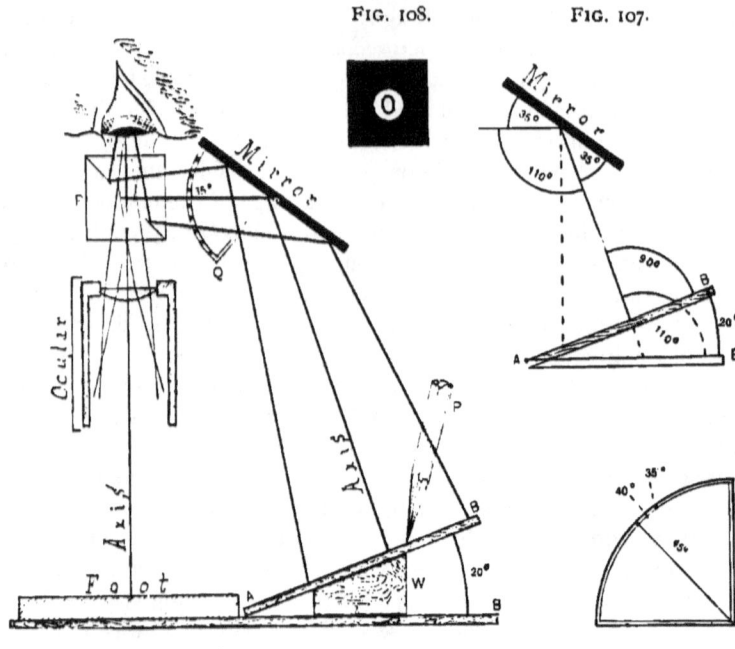

FIG. 106.   FIG 109.

FIGS. 106-109. *Abbe Camera Lucida in position to avoid distortion.*
FIG. 106. *The Abbe Camera Lucida with the mirror at 35°.*
Axis, Axis. Axial ray from the microscope and from the drawing surface.
A B. Drawing surface raised toward the microscope 20°.
Foot. The foot or base of the microscope.
Mirror with quadrant (Q). The mirror is seen to be at an angle of 35°.
Ocular. Ocular of the microscope.
P, P. Drawing pencil, and the cubical prism over the ocular.
W. Wedge to support the drawing board.

FIG. 107. *Geometrical figure of the preceding, showing the angles made by the axial ray with the mirror and the necessary elevation of the drawing board to avoid distortion. From the equality of opposite angles, the angle of the axial ray reflected at 35° must make an angle of 110° with a horizontal drawing board. The board must then be elevated toward the microscope 20° in order that the axial ray may be perpendicular to it, and thus fulfill the requirements necessary to avoid distortion (§ 170, 174).*

FIG. 108. *Upper view of the prism of the camera lucida. A considerable portion of the face of the prism is covered, and the opening in the silvered surface appears oval.*

FIG. 109. *Quadrant to be attached to the mirror of the Abbe Camera Lucida to indicate the angle of the mirror. As the angle is nearly always at 45°, 40°, or 35°, only those angles are shown.*

known angle in a moment, then the drawing board can be hinged and the elevation of 10° and 20° determined with a protractor. The drawing board is very conveniently held up by a broad wedge. By marking the position of the wedge for 10° and 20° the protractor need be used but once, then the wedge may be put into position at any time for the proper elevation.

§ 175. **Abbe Camera and Inclined Microscope.**—It is very fatiguing to draw continuously with a vertical microscope, and many mounted objects admit of an inclination of the microscope, when one can sit and work in a more comfortable position. The Abbe camera is as perfectly adapted to use with an inclined as with a vertical microscope. All that is requisite is to be sure that the fundamental law is observed regarding the axial ray of the image and the drawing surface, viz., that they should be at right angles. This is very easily accomplished as follows: The drawing board is raised toward the microscope twice as many degrees as the mirror is depressed below 45° (§ 174), then it is raised exactly as many degrees as the microscope is inclined, and in the same direction, that is, so the end of the drawing board shall be in a plane parallel with the stage of the microscope. The mirror must have its edges in planes parallel with the edges of the drawing board also (Fig. 110).

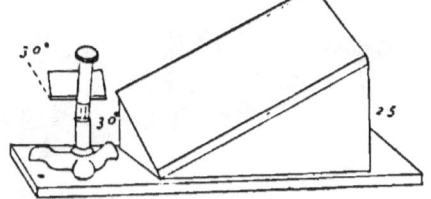

FIG. 110. *Arrangement of the drawing board for using the microscope in an inclined position with the Abbe camera lucida (designed by Mrs. S. P. Gage).*

A very elaborate and convenient drawing board has been devised by Bernhard (Zeit. wiss. Mikroskopie, Vol. XI, (1894) p. 298), whereby the proper inclination can be given the drawing board for the vertical microscope and also for an inclined microscope. The drawing surface as a whole can be raised or lowered to meet the needs of different objects. Fig. 111 shows an excellent drawing board after the Bernhard form.

§ 176. **Drawing with the Abbe Camera Lucida.**—(A) The light from the microscope and from the drawing surface should be of nearly equal intensity, so that the image and the drawing pencil can be seen with about equal distinctness,. This may be accomplished with very low powers (16 mm. and lower objectives) by covering the mirror with white paper when transparent objects are to be drawn. For high powers it is best to use a substage condenser. Often the light may be balanced by using a larger or smaller opening in the diaphragm. One can tell which field is excessively illuminated, for it is the one in which objects are most distinctly seen. If it is the microscopic, then the image of the microscopic object is very distinct and the pencil is invisible or very indistinct. If the drawing surface is too brilliantly lighted the pencil can be seen clearly, but the microscopic image will be very obscure.

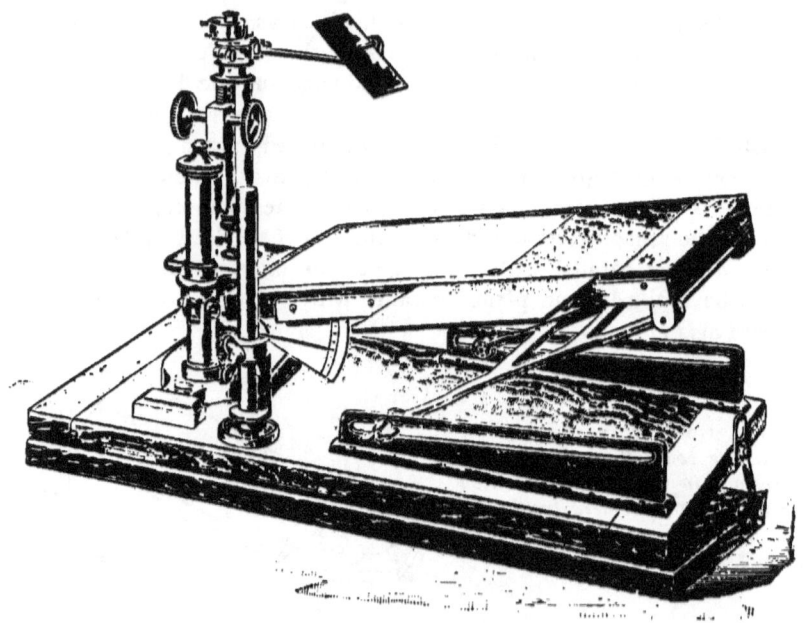

FIG. 111. *Drawing Board for the Abbe Camera Lucida* This drawing board, devised by the Bausch & Lomb Optical Co., is adjustable vertically, and the board may be inclined to prevent distortion. It is also arranged for use with an inclined microscope by having the base board hinged. Microscope and drawing surface are then inclined together. The camera lucida has a graduated arm to bear the mirror and a graduated quadrant at the mirror joint so that the angle of the mirror may be accurately determined. (See also Fig. 105). (From the Bausch & Lomb Optical Co.)

When opaque objects, that is objects which must be lighted with reflected light (§ 59), like dark colored insects, etc., are to be drawn the light must usually be concentrated upon the object in some way. The microscope may be placed in a very strong light and the drawing board shaded or the light may be concentrated upon the object by means of a concave mirror or a bull's eye condenser (Fig. 52).

If the drawing surface is too brilliantly illuminated, it may be shaded by placing a book or a ground glass screen between it and the window, also by putting one or more smoked glasses in the path of the rays from the mirror (Fig. 102 G). If the light in the microscope is too intense, it may be lessened by using white paper over the mirror, or by a ground glass screen between the microscope mirror and the source of light (Piersol, Amer. M. M. Jour., 1888, p. 103). It is also an excellent plan to blacken the end of the drawing pencil with carbon ink. Sometimes it is easier to draw on a black surface, using a white pencil or style. The carbon paper used in manifolding letters, etc., may be used, or ordinary black paper may be lightly rubbed on one side with a moderately soft lead pencil. Place the black paper over white paper and trace the outlines with a pointed style of ivory or bone. A corresponding dark line will appear on the white paper beneath. (Jour. Roy. Micr. Soc., 1883, p. 423).

(A) It is desirable to have the drawing paper fastened with thumb tacks, or in some other way. (B) The lines made while using the camera lucida should be very light, as they are liable to be irregular. (C) Only outlines are drawn and parts located with a camera lucida. Details are put in free-hand. (D) It is sometimes desirable to draw the outline of an object with a moderate power and add the details with a higher power. If this is done it should always be clearly stated. It is advisable to do this only with objects in which the same structure is many times duplicated, as a nerve or a muscle. In such an object all the different structures could be shown, and by omitting some of the fibers the others could be made plainer without an undesirable enlargement of the entire figure.

(E) If a drawing of a given size is desired and it cannot be obtained by any combination of oculars, objectives and lengths of the tube of the microscope, the distance between the camera lucida and the table may be increased or diminished until the image is of the desired size. This distance is easily changed by the use of a book or a block, but more conveniently if one has a drawing board with adjustable drawing surface like that shown in Fig. 111. The image of a few spaces of the micrometer, will give the scale of enlargement, or the power may be determined for the special case (§ 177-178).

(F) It is of the greatest advantage, as suggested by Heinsius Zeit. w. Mikr., 1889, p. 367), to have the camera lucida hinged so that the prism may be turned off the ocular for a moment's glance at the preparation, and then returned in place without the necessity of loosening screws and readjusting the camera. This form is now made by several opticians, and the quadrant is added by some. Any skilled mechanic can add the quadrant.

§ 177. **Magnification of the Microscope and Size of Drawings with the Abbe Camera Lucida.**—In determining the standard distance of 250 millimeters at which to measure the image in getting the magnification of the microscope, it is necessary to measure from the point marked P on the prism (Fig. 102) to the axis of the mirror and then vertically to the drawing board.

In getting the scale to which a drawing is enlarged the best way is to remove the preparation and put in its place a stage micrometer, and to trace a few (5 or 10) of its lines upon one corner of the drawing. The value of the spaces of the micrometer being given, thus,

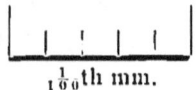

$\frac{1}{100}$th mm.

FIG. 112. *Showing the method of indicating the scale at which a drawing was made.*

The enlargement of the figure can then be accurately determined at any time by measuring with a steel scale the length of the image of the micrometer spaces and dividing it by their known width.

Thus, suppose the 5 spaces of the scale of enlargement given with a drawing were found to measure 25 millimeters and the spaces on the micrometer were $\frac{1}{100}$th millimeter, then the enlargement would be $25 \div \frac{5}{100} = 500$. That is, the image was drawn at a magnification of 500 diameters.

If the micrometer scale is used with every drawing, there is no need of troubling one's self about the exact distance at which the drawing is made, convenience may settle that, as the special magnification in each case may be determined from the scale accompanying the picture. It should be remembered, however, that the conditions when the scale is drawn must be exactly as when the drawing was made.

§ 178. **Drawing at Slight Magnification.**—Some objects are of considerable size and for drawings should be enlarged but a few diameters,—5 to 20. By using sufficiently low objectives and different oculars a great range may be obtained. Frequently, however, the range must be still further increased. For a moderate increase in size the drawing surface may be put farther off, or, as one more commonly needs less rather than greater magnification, the drawing surface may be brought nearer the mirror of the camera lucida by piling books or other objects on the drawing board. If one takes the precaution to draw a scale on the figure under the same conditions, its enlargement can be readily determined (§ 177).

If one has many large objects to draw at a low magnification, then some form of embryograph is very convenient. The writer has made use of a photographic camera and different photographic objectives for the purpose. The object is illuminated as if for a photograph and in place of the ground glass a plain glass is used and on this some tracing paper is stretched. Nothing is then easier than to trace the outlines of the object. See also Ch. VIII.

REFERENCES.

Beale, 31, 355; Behrens, Kossel and Schiefferdecker, 77; Carpenter-Dallinger, 233; Van Heurck, 91; American Naturalist, 1886, p. 1071, 1887, pp. 1040-1043; Amer. Monthly Micr. Jour., 1888, p. 103, 1890, p. 94; Jour. Roy. Micr. Soc., 1881, p. 819, 1882, p. 402, 1883, pp. 283, 560, 1884, p. 115, 1886, p. 516, 1888, pp. 113, 809, 798; Zeit. wiss. Mikroskopie, 1884, pp. 1-21, 1889, p. 367. 1893, pp. 289-295. Here is described an excellent apparatus made by Winkel. See Zeiss' catalog No 30, and the 15th (1896) edition of the Bausch & Lomb Optical Company for improved forms of the Abbe camera lucida and for improved drawing boards to accompany it.

CHAPTER VI.

MICRO-SPECTROSCOPE AND POLARISCOPE.

APPARATUS AND MATERIAL REQUIRED FOR THIS CHAPTER.

Compound microscope; Micro-spectroscope (§ 179); Watch-glasses and small vials, slides and covers (§ 198); Various substances for examination (as blood and ammonium sulphide, permanganate of potash, chlorophyll, some colored fruit, etc., (§ 199-209); Micro-polarizer (§ 211); Selenite plate (§ 220); Various doubly refracting objects, as crystals, textile fibers, starch, section of bone, etc.

MICRO-SPECTROSCOPE.

§ 179. **A Micro-Spectroscope, Spectroscopic or Spectral Ocular**, is a direct vision spectroscope in connection with a microscopic ocular. The one devised by Abbe and made by Zeiss consists of a direct vision spectroscope prism of the Amici pattern, and of considerable dispersion, placed over the ocular of the microscope. This direct vision or Amici prism consists of a single triangular prism of heavy flint glass in the middle and one of crown glass on each side, the edge of the crown glass prisms pointing toward the base of the flint glass prism, *i. e.*, the edges of the crown and flint glass prisms point in opposite directions. The flint glass prism serves to give the dispersion or separation into colors, while the crown glass prisms serve to make the emergent rays approximately parallel with the incident rays, so that one looks directly into the prism along the axis of the microscope.

The Amici prism is in a special tube which is hinged to the ocular and held in position by a spring. It may be swung free of the ocular. In connection with the ocular is the slit mechanism and a prism for reflecting horizontal rays vertically for the purpose of obtaining a comparison spectrum (§ 192). Finally near the top is a lateral tube with mirror for the purpose of projecting an Ångström scale of wave lengths upon the spectrum (§ 193, Figs. 113-114).

§ 180. **Apparent Reversal of the Position of the Colors in a Direct Vision Spectroscope.**—In accordance with the statements in § 179 the dispersion or separation into colors is given by the flint glass prism or prisms and in accordance with the general law that the waves of shortest length, blue, etc., will be bent most, consequently the colors have the position indicated in the top of Fig. 117, also above Fig. 118. But if one looks into the direct vision spectroscope or holds the eye close to the single prism (Fig. 118), the colors will appear reversed as if the red were more bent. The explanation of this is shown in Fig. 118, where it can be readily seen that if the eye is placed at E, close to the prism, the different colored rays

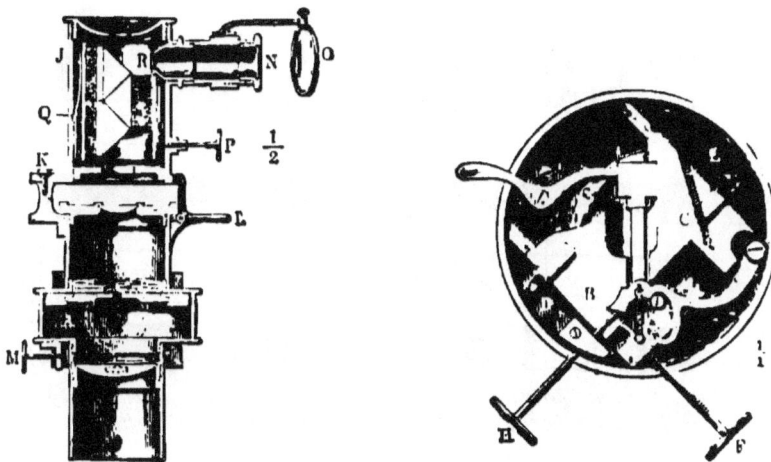

Fig. 113.
Longitudinal Section of the whole instrument.
(½ Full Size.)

Abbe's Micro-spectroscope.

Fig. 114.
Slit Mechanism separately.
(Plan view, Full size.)

"The eye lens is adjustable so as to accurately focus on the slit situated between the lenses. The mechanism for contracting and expanding the slit is actuated by the screw F and causes the laminae to move symmetrically (Merz's movement The slit may be made sufficiently wide so as to include the whole visual field. The screw H serves to limit the length of the slit so as to completely fill the latter with the image of the object under investigation when the comparison prism is inserted. The comparison prism is provided with a lateral frame and clips to hold the object and the illuminating mirror. All these parts together with the eye-piece are encased in a drum.

Above the eye-piece is placed an Amici prism of great dispersion which may be turned aside about the pivot K, so as to allow of the adjustment of the object being controlled, the prism being retained in its axial position by the spring catch L. A scale is projected on the spectrum by means of a scale tube and mirror attached to the prism casing. The divisions of the scale indicate in decimals of a micron the wave length of the respective section of the spectrum. The screw P serves to adjust the scale relative to the spectrum.

The instrument is inserted in the tube in place of the ordinary eye-piece and is clamped to the former by means of the screw M in such a position that the mirrors A and O, which respectively serve to illuminate the comparison prisms and the scale of wave lengths, are simultaneously illuminated." (Zeiss Catalog, No. 30.)

will appear in the direction from which they reach the eye and consequently are crossed in being projected into the field of vision and the real position is inverted. The same is true in looking into the micro-spectroscope. The actual position of the different colors may be determined by placing some ground glass or some of the lens-paper near the prism and observing with the eye at the distance of distinct vision.*

* The author wishes to acknowledge the aid rendered by Professor E. L. Nichols in giving the explanation offered in this section.

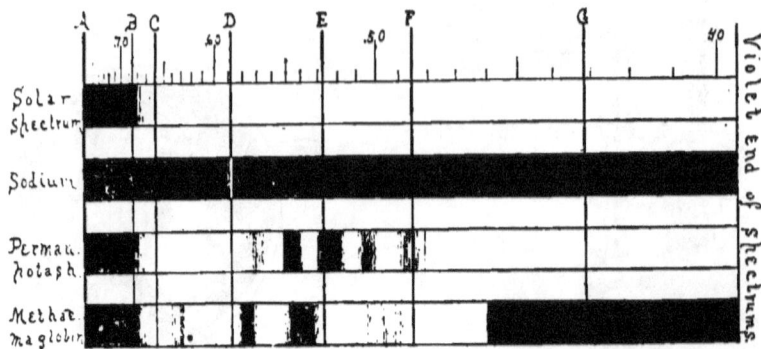

FIG. 115. *Various Spectrums.—All except that of Sodium were obtained by diffused day-light with the slit of such a width as gave the most distinct Fraunhofer lines.*

It frequently occurs that with a substance giving several absorption bands (e. g., chlorophyll) the density or thickness of the solution must be varied to show all the different bands clearly.

Solar Spectrum.— With diffused day-light and a narrow slit the spectrum is not visible much beyond the fixed line B. In order to extend the visible spectrum in the red to the line A, one should use direct sunlight and a piece of ruby glass in place of the watch glass in Fig. 117.

Sodium Spectrum.—The line spectrum (¿ 182) of sodium obtained by lighting the microscope with an alcohol flame in which some salt of sodium is glowing. With the micro-spectroscope the sodium line seen in the solar spectrum and with the incandescent sodium appears single, except under very favorable circumstances (¿ 193). By using a comparison spectrum of day-light with the sodium spectrum the light and dark D-lines will be seen to be continuous as here shown.

Permanganate of Potash.—This spectrum is characterized by the presence of five absorption bands in the middle of the spectrum and is best shown by using a $\frac{1}{16}$ per cent. solution of permanganate in water in a watch glass as in Fig. 117.

Met-hemoglobin.—The absorption spectrum of met-hemoglobin is characterized by a considerable darkening of the blue end of the spectrum and of four absorption bands, one in the red near the line C and two between D and E nearly in the place of the two bands of oxy-hemoglobin; finally there is a somewhat faint, wide band near F. Such a met-hemoglobin spectrum is best obtained by making a solution of blood in water of such a concentration that the two oxy-hemoglobin bands run together (¿ 202), and then adding three or four drops of a $\frac{1}{16}$ per cent. aqueous solution of permanganate of potash or a few drops of hydrogen dioxid (H_2O_2). Soon the bright red will change to a brownish color, when it may be examined.

VARIOUS KINDS OF SPECTRA.

By a spectrum is meant the colored bands appearing when light traverses a dispersing prism or a diffraction grating, or is affected in any way to separate the different wave lengths of light into groups. When daylight or some good artificial light is thus dispersed one gets the appearance so familiar in the rainbow.

§ 181. **Continuous Spectrum.**—In case a good artificial light or the electric light is used the various rainbow or spectral colors merge gradually into one another in passing from end to end of the spectrum. There are no breaks or gaps.

§ 182. **Line Spectrum.**—If a gas is made incandescent, the spectrum it produces consists, not of the various rainbow colors, but of sharp, narrow, bright lines, the color depending on the substance. All the rest of the spectrum is dark. These line spectra are very strikingly shown by various metals heated till they are in the form of incandescent vapor.

§ 183. **Absorption Spectrum.**—By this is meant a spectrum in which there are dark lines or bands in the spectrum. The most striking and interesting of the absorption spectra is the *Solar Spectrum*, or spectrum of sunlight. If this is examined carefully it will be found to be crossed by dark lines, the appearance being as if one were to draw pen marks across a continuous spectrum at various levels, sometimes apparently between the colors and sometimes in the midst of a color. These dark lines are the so called Fraunhofer Lines. Some of the principal ones have been lettered with Roman capitals, A, B, C, D, E, F, G, H, commencing at the red end. The meaning of these lines was for a long time enigmatical, but it is now known that they correspond with the bright lines of a line spectrum (§ 182). For example, if sodium is put in the flame of a spirit lamp it will vaporize and become luminous. If this light is examined there will be seen one or two bright yellow bands corresponding in position with D of the solar spectrum (Fig. 114). If now the spirit lamp-flame, colored by the incandescent sodium, is placed in the path of the electric light, and it is examined as before, there will be a continuous spectrum, except for dark lines in place of the bright sodium lines. That is, the comparatively cool yellow light of the spirit lamp cuts off or absorbs the intensely hot yellow light of the electric light; and although the spirit flame sends a yellow light to the spectroscope it is so faint in comparison with the electric light that the sodium lines appear dark. It is believed that in the sun's atmosphere there are incandescent metal vapors (sodium, iron, etc.), but that they are so cool in comparison with the rays of their wave length in the sun that the cooler light of the incandescent metallic vapors absorb the light of corresponding wave length, and are, like the spirit lamp flame, unable to make up the loss, and therefore the presence of the dark lines.

§ 184. **Absorption Spectra from Colored Substances.**—While the solar spectrum is an absorption spectrum, the term is more commonly applied to the spectra obtained with light which has passed through or has been reflected from colored objects which are not self-luminous.

It is the special purpose of the micro-spectroscope to investigate the spectra of colored objects which are not self-luminous, as blood and other liquids, various minerals, as monazite, etc. The spectra obtained by examining the light reflected from these colored bodies or transmitted through them, possess, like the solar spectrum dark lines or bands, but the bands are usually much wider and less sharply defined. Their number and position depend on the substance or its constitution (Fig. 116), and their width, in part, upon the thickness of the body. With some colored bodies, no definite bands are present. The spectrum is simply restricted at one or both ends and various of the other colors are considerably lessened in intensity. This is true of many colored fruits.

§ 185. **Angström and Stokes' Law of Absorption Spectra.** The wave lengths of light absorbed by a body when light is transmitted through some of its substance

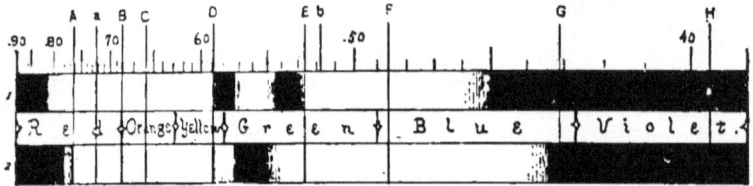

FIG. 116. *Absorption Spectrum of Oxy-Hemoglobin or arterial blood (1) and of Hemoglobin or venous blood (2). (From Gamgee and McMunn.)*
A, B, C, D, E, F, G, H. Some of the Principal Fraunhofer lines of the solar spectrum (§ 183).
.90, .80, .70, .60, .50, .40. Wave lengths in microns, as shown in Angström's scale (§ 193). It will be seen that the wave lengths increase toward the red and decrease toward the violet end of the spectrum.
Red, Orange, Yellow, etc. Color regions of the spectrum. Indigo should come between the blue and the violet to complete the seven colors usually given. It was omitted through inadvertence.

are precisely the waves radiated from it when it becomes self-luminous. For example, a piece of glass that is yellow when cool, gives out blue light when it is hot enough to be self-luminous. Sodium vapor absorbs two bands of yellow light (D lines); but when light is not sent through it, but itself is luminous and examined as a source of light its spectrum gives bright sodium lines, all the rest of the spectrum being dark.

§ 186. **Law of Color.**—The light reaching the eye from a colored, solid, liquid or gaseous body lighted with white light, will be that due to white light less the light waves that have been absorbed by the colored body. Or in other words, it will be due to the wave lengths of light that finally reach the eye from the object. For example, a thin layer of blood under the microscope will appear yellowish green, but a thick layer will appear pure red. If now these two layers are examined with a micro-spectroscope, the thin layer will show all the colors, but the red end will be slightly, and the blue end considerably restricted, and some of the colors will appear of considerably lessened intensity. Finally there may appear two shadow-like bands, or if the layer is thick enough, two well-defined dark bands in the green (§ 202).

If the thick layer is examined in the same way, the spectrum will show only red with a little orange light, all the rest being absorbed. Thus the spectroscope shows which colors remain, in part or wholly, and it is the mixture of this remaining or unabsorbed light that gives color to the object.

§ 187. **Complementary Spectra.**—While it is believed that Angström's law (§ 185) is correct, there are many bodies on which it cannot be tested, as they change in chemical or molecular constitution before reaching a sufficiently high temperature to become luminous. There are compounds, however, like those of didymium, erbium and terbium, which do not change with the heat necessary to render them luminous, and with them the incandescence and absorption spectra are mutually complementary, the one presenting bright lines where the other presents dark ones (Daniell).

ADJUSTING THE MICRO-SPECTROSCOPE.

§ 188. The micro-spectroscope, or spectroscopic ocular, is put in the place of the ordinary ocular in the microscope, and clamped to the top of the tube by means of a screw for the purpose.

§ 189. **Adjustment of the Slit.**—In place of the ordinary diaphragm with circular opening, the spectral ocular has a diaphragm composed of two movable knife edges by which a slit-like opening of greater or less width and length may be obtained at will by the use of screws for the purpose. To adjust the slit, depress the lever holding the prism-tube in position over the ocular, and swing the prism aside. One can then look into the ocular. The lateral screw should be used and the knife edges approach till they appear about half a millimeter apart. If now the Amici prism is put back in place and the microscope well lighted, one will see a spectrum by looking into the upper end of the spectroscope. If the slit is too wide, the colors will overlap in the middle of the spectrum and be pure only at the red and blue ends; and the Fraunhofer or other bands in the spectrum will be faint or invisible. Dust on the edges of the slit gives the appearance of longitudinal streaks on the spectrum.

§ 190. **Mutual Arrangement of Slit and Prism.**—In order that the spectrum may appear as if made up of colored bands going directly across the long axis of the spectrum, the slit must be parallel with the refracting edge of the prism. If the slit and prism are not thus mutually arranged, the colored bands will appear oblique, and the whole spectrum may be greatly narrowed. If the colored bands are oblique, grasp the prism tube and slowly rotate it to the right or to the left until the various colored bands extend directly across the spectrum.

§ 191. **Focusing the Slit.**—In order that the lines or bands in the spectrum shall be sharply defined, the eye-lens of the ocular should be accurately focused on the slit. The eye-lens is movable, and when the prism is swung aside it is very easy to focus the slit as one focused for the ocular micrometer (§ 161). If one now uses daylight there will be seen in the spectrum the dark Fraunhofer lines (Fig. 116 E. F., etc.

To show the necessity of focusing the slit, move the eye-lens down or up as far as possible, and the Fraunhofer lines cannot be seen. While looking into the spectroscope move the ocular lens up or down, and when it is focused the Fraunhofer lines will reappear. As the different colors of the spectrum have different wave lengths, it is necessary to focus the slit for each color if the sharpest possible pictures are desired.

It will be found that the eye-lens of the ocular must be farther from

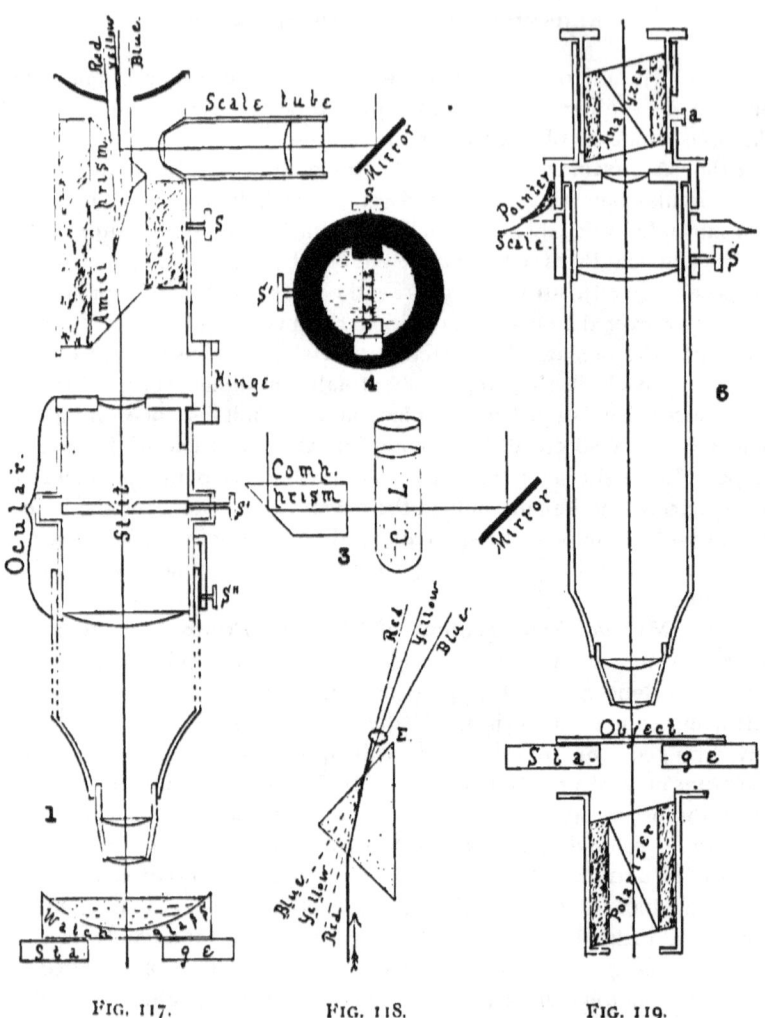

Fig. 117.   Fig. 118.   Fig. 119.

Fig. 117 (1). *Section of the tube and stage of the microscope with the spectral ocular or micro-spectroscope in position.*

Amici Prism (½ 167).—*The direct vision prism of Amici in which the central shaded prism of flint glass gives the dispersion or separation into colors, while the end prisms of crown glass cause the rays to emerge approximately parallel with the axis of the microscope. A single ray is represented as entering the prism and this is divided into three groups (Red, Yellow, Blue), which emerge from the prism, the red being least and the blue most bent toward the base of the flint prism* (see Fig. 118).

Hinge.—The hinge on which the prism tube turns when it is swung off the ocular.

Ocular (§ 179) —*The ocular in which the slit mechanism takes the place of the diaphragm* (§ 189). *The eye-lens is movable as in a micrometer ocular, so that the slit may be accurately focused for the different colors* (§ 191).

S. Screw for setting the scale of wave lengths (§ 193).

S'. Screw for regulating the width of the slit (§ 189).

S''. Screw for clamping the micro-spectroscope to the tube of the microscope.

Scale Tube.—The tube near the upper end containing the Ångström scale and the lenses for projecting the image upon the upper face of the Amici prism, whence it is reflected upward to the eye with the different colored rays. At the right is a special mirror for lighting the scale. For arranging and focusing the scale, (see § 193).

Slit.—*The linear opening between the knife edges* Through the slit the light passes to the prism. It must be arranged parallel with the refracting edge of the prism, and of such a width that the Fraunhofer or Fixed Lines are very clearly and sharply defined when the eye-lens is properly focused (§ 189-191).

Stage.—*The stage of the microscope. This supports a watch-glass with sloping sides for containing the colored liquid to be examined.*

(3) *Comparison Prism with tube for colored liquid* (*C. L.*), *and mirror.* The prism reflects horizontal rays vertically, so that when the prism is made to cover part of the slit two parallel spectra may be seen, one from light sent directly through the entire microscope and one from the light reflected upward from the comparison prism.

(4) *View of the Slit Mechanism from below.*—*Slit, the linear space between the knife edges through which the light passes.*

P. Comparison prism beneath the slit and covering part of it at will.

S. S'. Screws for regulating the width and length of the slit.

Fig. 118. *Flint-Glass Prism showing the separation or dispersion of white light into the three groups of colored rays* (*Red, Yellow, Blue*), *the blue rays being bent the most from the refracting edge* (§ 180).

Fig. 119. *Sectional View of a Microscope with the Polariscope in Position* (§ 209-217).

Analyzer and Polarizer.—*They are represented with corresponding faces parallel so that the polarized beam could traverse freely the analyzer. If either nicol were rotated 90° they would be crossed and no light would traverse the analyzer unless some polarizing substance were used as object* (§ 212). (a) *Slot in the analyzer tube so that the analyzer may be raised or lowered to adjust it for difference of level of the eye-point in different oculars* (§ 214).

Pointer and Scale.—The pointer attached to the analyzer and the scale on divided circle clamped (by the screw *S*) to the tube of the microscope. The pointer and scale enable one to determine the exact amount of rotation of the analyzer 211.

Object.—*The object whose character is to be investigated by polarized light.*

the slit for the sharpest focus of the red end than for the sharpest focus of the lines at the blue end. This is because the wave length of red is markedly greater than for blue light.

Longitudinal dark lines on the spectrum may be due to irregularity of the edge of the slit or to the presence of dust. They are most troublesome with a very narrow slit.

§ 192. **Comparison or Double Spectrum.**—In order to compare the spectra of two different substances it is desirable to be able to examine their spectra side by side. This is provided for in the better forms of micro-spectroscopes by a prism just below the slit, so placed that the light entering it from a mirror at the side of the drum shall be totally reflected in a vertical direction, and thus parallel with the rays from the microscope. The two spectra will be side by side with a narrow dark line separating them. If now the slit is well focused and daylight be sent through the microscope and into the side to the reflecting or comparison prism, the colored bands and the Fraunhofer dark lines will appear directly continuous across the two spectra. The prism for the comparison spectrum is movable and may be thrown entirely out of the field if desired. When it is to be used, it is moved about half way across the field so that the two spectrums shall have about the same width.

§ 193. **Scale of Wave Lengths.**—In the Abbe micro-spectroscope the scale is in a separate tube near the top of the prism and at right angles to the prism-tube. A special mirror serves to light the scale, which is projected upon the spectrum by a lens in the scale-tube. This scale is of the Ångström form, and the wave lengths of any part of the spectrum may be read off directly, after the scale is once set in the proper position, that is, when it is set so that any given wave length on the scale is opposite the part of the spectrum known by previous investigation to have that particular wave length. The point most often selected for setting the scale is opposite the sodium lines where the wave length is, according to Ångström, $0.5892\ \mu$. In adjusting the scale, one may focus very sharply the dark sodium line of the solar spectrum and set the scale so that the number 0.589 is opposite the sodium or D line, or a method that is frequently used and serves to illustrate § 171, is to sprinkle some salt of sodium (carbonate of sodium is good) in an alcohol lamp flame and to examine this flame. If this is done in a darkened place with a spectroscope, a narrow bright band will be seen in the yellow part of the spectrum. If now ordinary daylight is sent through the comparison prism, the bright line of the sodium will be seen to be directly continuous with the dark line at D in the solar spectrum (Fig. 114). Now, by reflecting light into the scale-tube the image of the scale will appear on the spectrum, and by a screw just under the scale-tube, but in the prism-tube, the proper point on the scale (0.589 μ) can be brought opposite the sodium band. All the scale will then give the wave lengths directly. Sometimes the scale is oblique to the spectrum. This may be remedied by turning the prism-tube slightly one way or the other. It may be due to the wrong position of the scale itself. If so, grasp the milled ring at the distal end of the scale-tube and, while looking into the spectroscope, rotate the tube until the lines of the scale are parallel with the Fraunhofer lines. It is necessary in adjusting the scale to be sure that the larger number, 0.70, is at the red end of the spectrum.

The numbers on the scale should be very clearly defined. If they do not so appear, the scale-tube must be focused by grasping the outer tube of the scale-tube and moving it toward or from the prism-tube until the scale is distinct. In focusing the scale, grasp the outer scale-tube with one hand and the prism-tube with the other, and push or pull in opposite directions. In this way one will be less liable to injure the spectroscope.

§ 194. **Designation of Wave Length.**—Wave lengths of light are designated by the Greek letter λ, followed by the number indicating the wave length in some fraction of a meter. With the Abbe micro-spectroscope the micron is taken as the unit as with other microscopical measurements (§ 157). Various units are in use, as the one hundred thousandth of a millimeter, millionths or ten millionths of a millimeter. If these smaller units are taken, the wave lengths will be indicated either as a decimal fraction of a millimeter or as whole numbers. Thus, according to Angström, the wave length of sodium light is 5892 ten millionths mm., or 589.2 millionths, or 58.92 one hundred thousandths, or 0.5892 one thousandth mm., or 0.5892 μ. The last would be indicated thus, λ D = 0.5892 μ.

§ 195. **Lighting for the Micro-spectroscope.**—For opaque objects a strong light should be thrown on them either with a concave mirror or a condensing lens. For transparent objects the amount of the substance and the depth of color must be considered. As a general rule it is well to use plenty of light, as that from an Abbe illuminator with a large opening in the diaphragm, or with the diaphragm entirely removed. For very small objects and thin layers of liquids it may be better to use less light. One must try both methods in a given case, and learn by experience.

The direct and the comparison spectrums should be about equally illuminated. One can manage this by putting the object requiring the greater amount of illumination on the stage of the microscope and lighting it with the Abbe illuminator. In lighting it is found in general that for red or yellow objects, lamp-light gives very satisfactory results. For the examination of blood and blood crystals, the light from a petroleum lamp is excellent (§ 201-203). For objects with much blue or violet, daylight or artificial light rich in blue light is best. The new acetylene light ought to be very satisfactory (§ 65).

Furthermore, one should be on his guard against confusing the ordinary absorption bands with the Fraunhofer lines when daylight is used. With lamp-light the Fraunhofer lines are absent and, therefore, not a source of possible confusion.

§ 196. **Objectives to Use with the Micro-spectroscope.**—If the material is of considerable bulk, a low objective (18 to 50 mm.) is to be preferred. This depends on the nature of the object under examination, however. In case of individual crystals one should use sufficient magnification to make the real image of the crystal entirely fill the width of the slit. The length of the slit may then be regulated by the screw on the side of the drum, and also by the comparison prism. If the object does not fill the whole slit the white light entering the spectroscope with the light from the object might obscure the absorption bands.

In using high objectives with the micro-spectroscope one must very carefully regulate the light (§ 58, 102), and sometimes shade the object.

§ 197. **Focusing the Objective.**—For focusing the objective the prism-tube is swung aside, and then the slit made wide by turning the adjusting screw at the side. When the slit is open, one can see objects when the microscope is focused as with an ordinary ocular. After an object is focused, it may be put exactly in position to fill the slit of the spectroscope, then the knife edges are brought together till the slit is of the right width ; if the slit is then too long it may be shortened by using one of the mechanism screws on the side, or if that is not sufficient, by bringing the comparison prism farther over the field. If one now replaces the Amici prism and looks into the microscope, the spectrum is liable to have longitudinal shimmering lines. To get rid of these focus up or down a little so that the microscope will be slightly out of focus.

§ 198. **Amount of Material Necessary for Absorption Spectra and its Proper Manipulation.**—The amount of material necessary to give an absorption spectrum varies greatly with different substances, and can be determined only by trial. If a transparent solid is under investigation it is well to have it in the form of a wedge, then successive thicknesses can be brought under the microscope. If a liquid substance is being examined, a watch glass with sloping sides forms an excellent vessel to contain it, then successive thicknesses of the liquid can be brought into the field as with the wedge-shaped solid. Frequently only a very weak solution is obtainable; in this case it can be placed in a homœopathic vial, or in some glass tubing sealed at the end, then one can look lengthwise through the liquid and get the effect of a more concentrated solution. For minute bodies like crystals or blood corpuscles, one may proceed as described in the previous section.

MICRO-SPECTROSCOPE—EXPERIMENTS.

§ 199. Put the micro-spectroscope in position, arrange the slit and the Amici prism so that the spectrum will show the various spectral colors going directly across it (§ 188-189) and carefully focus the slit. This may be done either by swinging the prism-tube aside and proceeding as for the ocular micrometer (§ 161), or by moving the eye-lens of the ocular up and down while looking into the micro-spectroscope until the dark lines of the solar spectrum are distinct. If they cannot be made distinct by focusing the slit, then the light is too feeble or the slit is too wide (§ 191). With the lever move the comparison prism across half the field so that the two spectra shall be of about equal width. For lighting, see § 195.

§ 200. **Absorption Spectrum of Permanganate of Potash.**—Make a solution of permanganate of potash in water of such a strength that a stratum 3 or 4 mm. thick is transparent. Put this solution in a watch-glass with sloping sides, and put it under the microscope. Use a 50 mm. or 16 mm. objective, and use the full opening of the illuminator. Light strongly. Look into the spectroscope and slowly move the watch-glass into the field. Note carefully the appearance with the thin stratum of liquid at the edge and then as it gradually thickens on moving the watch-glass still farther along. Count the absorption bands and note particularly the red and blue ends. Compare carefully with the comparison spectrum (Fig. 113). For strength of solution see § 202.

§ 201. **Absorption Spectrum of Blood.**—Obtain blood from a recently killed animal, or flame a needle, and after it is cool prick the finger two or three times in a small area, then wind a handkerchief or a rubber tube around the base of the finger, and squeeze the finger with the other hand. Some blood will ooze out of the pricks. Rinse this off in a watch-glass partly filled with water. Continue to add the blood until the water is quite red. Place the watch-glass of diluted blood under the microscope in place of the permanganate, using the same objective, etc. Note carefully the spectrum. It would be advantageous to determine the wave length opposite the center of the dark bands. This may be done easily by setting the scale properly as described in § 193. Make another preparation, but use a homœopathic vial instead of a watch-glass. Cork the vial and lay it down upon the stage of the microscope. Observe the spectrum. It will be like that in the watch-glass. Remove the cork and look through the whole length of the vial. The bands will be very much darker, and if the solution is thick enough only red and a little orange will appear. Re-insert the cork and incline the vial so that the light traverses a very thin layer, then gradually elevate the vial and the effect of a thicker and thicker layer may be seen. Note especially that the two characteristic bands unite and form one wide band as the stratum of liquid thickens. Compare with the following:

Add to the vial of diluted blood a drop or two of ammonium sulphide, such as is used for a reducing agent in chemical laboratories. Shake the bottle gently and then allow it to stand for ten or fifteen minutes. Examine it and the two bands will have been replaced by a single, less clearly defined band in about the same position. The blood will also appear somewhat purple. Shake the vial vigorously and the color will change to the bright red of fresh blood. Examine it again with the spectroscope and the two bands will be visible. After five or ten minutes another examination will show but a single band. Incline the bottle so that a very thin stratum may be examined. Note that the stratum of liquid must be considerably thicker to show the absorption band than was necessary to show the two bands in the first experiment. Furthermore, while the single band may be made quite black on thickening the stratum, it will not separate into two bands with a thinner stratum. In this experiment it is very instructive to have a second vial of fresh diluted blood, say that from the watch-glass, before the opening of the comparison prism. The two banded spectrum will then be in position to be compared with the spectrum of the blood treated with the ammonium sulphide.

The two banded spectrum is of *oxy-hemoglobin*, or arterial blood, the single banded spectrum is of *hemoglobin* (sometimes called reduced hemoglobin) or venous blood, that is, the respiratory oxygen is present in the two banded spectrum but absent from the single banded spectrum. When the bottle was shaken the hemoglobin took up oxygen from the air and became oxy-hemoglobin, as occurs in the lungs, but soon the ammonium sulphide took away the respiratory oxygen, thus reducing the oxy-hemoglobin to hemoglobin. This may be repeated many times (Fig. 114).

§ 202. **Met-Hemoglobin.**—The absorption spectrum of met-hemoglobin is characterized by a considerable darkening of the blue end of the spectrum and of four absorption bands, one in the red near the line C and two between D and E, nearly in the place of the two bands of oxy-hemoglobin; finally there is a somewhat faint, wide band near F. Such a met-hemoglobin spectrum is best obtained by making a solution of blood in water of such a concentration that the two oxy-hemoglobin bands run together (§ 201), and then adding three or four drops of a $\frac{1}{10}$ per cent. aqueous solution of permanganate of potash. Soon the bright red will change to a brownish color, when it may be examined (Fig. 113).

§ 203. **Carbon Monoxide Hemoglobin** (CO Hemoglobin).—To obtain this one may kill an animal, after anæsthetization, in illuminating gas, or one may allow illuminating gas to bubble through some blood already taken from the body. The gas should bubble through a minute or two. The oxygen will be displaced by carbon monoxide. This forms quite a stable compound with hemoglobin, and is of a bright cherry-red color. Its spectrum is nearly like that of oxy-hemoglobin, but the bands are farther toward the blue. Add several drops of ammonium sulphide and allow the blood to stand some time. No reduction will take place, thus forming a marked contrast to solutions of oxy-hemoglobin. By the addition of a few drops of glacial acetic acid a dark brownish red color is produced.

§ 204. **Carmine Solution.**—Make a solution of carmine by putting $\frac{1}{10}$th gram of carmine in 100 cc. of water and adding 10 drops of strong ammonia. Put some of this in a watch-glass or in a small vial and compare the spectrum with that of oxy-hemoglobin or carbon monoxide hemoglobin. It has two bands nearly in the same position, thus giving the spectrum a striking similarity to blood. If now several drops, 15 or 20, of glacial acetic acid are added to the carmine, the bands remain and the color is not very markedly changed, while with either oxy-hemoglobin or CO-hemoglobin the color would be very markedly changed from the bright red to a dull reddish brown, and the spectrum, if any could be seen, would be markedly different. Carmine and O-hemoglobin can be distinguished by the use of ammonium sulphide, the carmine remaining practically unchanged while the blood shows the single band of hemoglobin (§ 201). The acetic acid serves to differentiate the CO-hemoglobin as well as the O-hemoglobin.

§ 205. **Colored Bodies not giving Distinctly Banded Absorp-**

tion Spectra.—Some quite brilliantly colored objects, like the skin of a red apple, do not give a banded spectrum. Take the skin of a red apple, mount it on a slide, put on a cover-lass and add a drop of water at the edge of the cover. Put the preparation under the microscope and observe the spectrum. Although no bands will appear, in some cases at least, yet the ends of the spectrum will be restricted and various regions of the spectrum will not be so bright as the comparison spectrum. Here the red color arises from the mixture of the unabsorbed wave lengths, as occurs with other colored objects. In this case, however, not all the light of a given wave length is absorbed, consequently there are no clearly defined dark bands, the light is simply less brilliant in certain regions and the red rays so predominate that they give the prevailing color.

§ 206. **Nearly Colorless Bodies with Clearly Marked Absorption Spectra.**—In contradistinction to the brightly colored objects with no distinct absorption bands are those nearly colorless bodies and solutions which give as sharply defined absorption bands as could be desired. The best examples of this are afforded by solutions of the rare earths, didymium, etc. These in solutions that give hardly a trace of color to the eye give absorption bands that almost rival the Fraunhofer lines in sharpness.

§ 207. **Absorption Spectra of Minerals.**—As example take some monazite sand on a slide and either mount it in balsam (see Ch. VII), or cover and add a drop of water. The examination may be made also with the dry sand, but it is less satisfactory. Light well with transmitted light, and move the preparation slowly around. Absorption bands will appear occasionally. Swing the prism-tube off the ocular, open the slit and focus the sand. Get the image of one or more grains directly in the slit, then narrow and shorten the slit so that no light can reach the spectroscope that has not traversed the grain of sand. The spectrum will be very satisfactory under such conditions. It is frequently of great service in determining the character of unknown mineral sands to compare their spectra with known minerals. If the absorption bands are identical, it is strong evidence in favor of the identity of the minerals. For proper lighting see § 195.

§ 208. While the study of absorption spectra gives one a great deal of accurate information, great caution must be exercised in drawing conclusions as to the identity or even the close relationship of bodies giving approximately the same absorption spectra. The rule followed by the best workers is to have a known body as control and to treat the unknown body and the known body with the same reagents, and to dissolve them in the same medium. If all the reactions are identical then the presumption is very strong that the bodies are identical or very closely related. For example, while one might be in doubt between a solution of oxy- or CO-hemoglobin and carmine, the addition of ammonium sulphide would serve to change the double to a single band in the O-hemoglobin, and glacial acetic acid would enable one to distinguish between the CO-blood and the carmine, although the ammonium sulphide would not enable one to make the distinction. Furthermore it is unsafe to compare objects dissolved in different media. The same objects as "cyanine and aniline blue dissolved in alcohol give a very similar spectrum, but in water a totally different one." "Totally different bodies show absorption bands in exactly the same position (solid nitrate of uranium and permanganate of potash in the blue)." (MacMunn). The rule given by MacMunn is a good one : "The recognition of a body becomes more certain if its spectrum consists of several absorption bands, but even the coincidence of these bands with those of another body, is not sufficient to enable us to infer chemical identity ; what enables us to do so with certainty is the fact : *that the two solutions give bands of equal intensities in the same parts of the spectrum which undergo analogous changes on the addition of the same reagent.*"

REFERENCES TO THE MICRO-SPECTROSCOPE AND SPECTRUM ANALYSIS.

The micro-spectroscope is playing an ever increasingly important role in the spectrum analysis of animal and vegetable pigments, and of colored mineral and chemical substances, therefore a somewhat extended reference to literature will be given. Full titles of the books and periodicals will be found in the Bibliography at the end.

Ångström, Recherches sur le spectre solaire, etc. Also various papers in periodicals. See Royal Soc's Cat'l Scientific Papers ; Anthony & Brackett ; Beale, p. 269; Behrens, p. 139; Kossel und Schiefferdecker, p. 63 ; Carpenter, p. 104 ; Browning, How to Work with the Spectroscope, and in Monthly Micr. Jour., II, p. 65 ; Daniell, Principles of Physics. The general principles of spectrum analysis are especially well stated in this work, pp. 435-455 ; Davis, p. 342 ; Dippel, p. 277 ; Frey ; Gamgee, p. 91 ; Halliburton ; Hogg, p. 122 ; also in Monthly Micr. Jour., Vol. II, on colors of flowers ; Jour. Roy. Micr. Soc., 1880, 1883, and in various other vols. ; Kraus ; Lockyer ; M'Kendrick ; MacMunn ; and also in Philos. Trans. R. S., 1886 ; various vols. of Jour. Physiol. ; Nägeli und Schwendener ; Proctor ; Ref. Hand-Book Med. Sciences, Vol. I, p. 577, VI, p. 516, VII, p. 426 ; Roscoe ; Schellen ; Sorby, in Beale, p. 269; also Proc. R. S., 1874, p. 31, 1867, p. 433 ; see also in the Scientific Review, Vol. V, p. 66, Vol. II, p. 419. The larger works on Physiology, Chemistry and Physics may also be consulted with profit.

Vogel, Spectrum analysis, also in Nature, Vol. xix, p. 495, on absorption spectra. The bibliography in MacMunn is excellent and extended.

MICRO-POLARISCOPE.

§ 209. The micro-polariscope, or polarizer, is a polariscope used in connection with a microscope.

The most common and typical form consists of two Nicol prisms, that is, two somewhat elongated rhombs of Iceland spar cut diagonally and cemented together with Canada balsam. These Nicol prisms are then mounted in such a way that the light passes through them lengthwise, and in passing is divided into two rays of plane polarized light. The one of these rays obeying most nearly the ordinary law of refraction is called the *ordinary ray*, the one departing farthest from the law is called the *extra-ordinary ray*. These two rays are not only polarized, but polarized in planes almost exactly at right angles to each other. The Nicol prism totally reflects the ordinary ray at the cemented surface as it meets that surface at an angle greater than the critical angle, and only the extraordinary or less refracted ray is transmitted.

§ 210. **Polarizer and Analyzer.**—The polarizer is one of the Nicol prisms. It is placed beneath the object and in this way the object is illuminated with polarized light. The analyzer is the other Nicol and is placed at some level above the object, very conveniently above the ocular.

When the corresponding faces of the polarizer and analyzer are parallel *i. e.*, when the faces through which the oblique section passed are parallel, light passes freely through the analyzer to the eye. If these corresponding faces are at right angles, that is, if the Nicols are crossed, then the light is entirely cut off and the two transparent prisms become opaque to ordinary light. There are then, in the complete revolution of the analyzer, two points, at 0° and 180°, where the corresponding faces are parallel and where light freely traverses the analyzer. There are also two crossing points of the Nicols, at 90° and 270°, where the light is extinguished. In the intermediate points there is a sort of twilight.

§ 211. **Putting the Polarizer and Analyzer in Position.**—Swing the diaphragm carrier of the Abbe illuminator out from under the illuminator, remove the disk diaphragm or open widely the iris diaphragm and place the analyzer in the diaphragm carrier, then swing it back under the illuminator. Remove the ocular, put the graduated ring on the top of the tube and then replace the ocular and put the analyzer over the ocular and ring. Arrange the graduated ring so that the indicator shall stand at 0° when the field is lightest. This may be done by turning the tube down so that the objective is near the illuminator, then shading the stage so that none but polarized light shall enter the microscope. Rotate the analyzer until the lightest possible point is found, then rotate the graduated ring till the index stands at 0°. The ring may then be clamped to the tube by the side screw for the purpose. Or, more easily, one may set the index at 0°, clamp the ring to the microscope, then rotate the draw-tube of the microscope till the field is lightest.

§ 212. **Adjustment of the Analyzer.**—The analyzer should be capable of moving up and down in its mounting, so that it can be adjusted to the eye-point of the ocular with which it is used. If on looking into the analyzer with parallel Nicols the edge of the field is not sharp, or if it is colored, the analyzer is not in a proper position with reference to the eye-point, and should be raised or lowered till the edge of the field is perfectly sharp and as free from color as the ocular with the analyzer removed.

§ 213. **Objectives to Use with the Polariscope.**—Objectives of the lowest power may be used, and also all intermediate forms up to a 2 mm. homogeneous immersion. Still higher objectives may be used if desired. In general, however, the lower powers are somewhat more satisfactory. A good rule to follow in this case is the general rule in all microscopic work,—*use the power that most clearly and satisfactorily shows the object under investigation.*

§ 214. **Lighting for Micro-Polariscope Work.**—Follow the general directions given in Chapter II. It is especially necessary to shade the object so that no unpolarized light can enter the objective, otherwise the field cannot be sufficiently darkened. No diaphragm is used over the polarizer for most examinations. Direct sunlight may be used to advantage with some objects, and as a rule the object would best be very transparent.

§ 215. **Mounting Objects for the Polariscope.**—So far as possible objects should be mounted in balsam to render them very transparent. In many cases objects mounted in water do not give satisfactory polariscopic appearances. For example, if starch is mounted dry or in water, the appearances are not so striking as in a balsam mount (Davis, p. 337 ; Suffolk).

§ 216. **Purpose of a Micro-Polariscope.**—The object of a micro-polariscope is to determine, in microscopic masses, one or more of the following points : (A) Whether the body is singly refractive, mono-refringent, or *isotropic*, that is, optically homogeneous, as are glass and crystals belonging to the cubical system ; (B) Whether the object is doubly refractive or *anisotropic*, uniaxial or biaxial ; (C) *Pleochroism ;* (D) The rotation of the plane of polarization, as with solutions of sugar, etc. ; (E) To aid in petrology and mineralogy ; (F) To aid in the determination of very minute quantities of crystallizable substances ; (G) For the production of colors.

For petrological and mineralogical investigations the microscope should possess a graduated, rotating stage so that the object can be rotated, and the exact angle of rotation determined. It is also found of advantage in investigating objects with polarized light where colors appear, to combine a polariscopic and spectroscope (Spectro-Polariscope).

MICRO-POLARISCOPE—EXPERIMENTS.

§ 217. Arrange the polarizer and analyzer as directed above (§ 211) and use a 16 mm. objective except when otherwise directed.

(A) **Isotropic or Singly Refractive Objects.**—Light the microscope well and cross the Nicols, shade the stage and make the field as dark as possible (§ 210). As an isotropic substance, put an ordinary glass slide under the microscope. The field will remain dark. As an example of a crystal belonging to the cubical system and hence isotropic, make a strong solution of common salt (sodium chloride Na Cl. , put a drop on a slide and allow it to crystallize, put it under the microscope, remove the analyzer, focus the crystals and then replace the analyzer and cross the Nicols. The field and the crystals will remain dark.

(B) **Anisotropic or Doubly Refracting Objects.**—Make a fresh preparation of carbonate of lime crystals like that described for pedesis (§ 142), or use a preparation in which the crystals have dried to the slide, use a 5 or 3 mm. objective, shade the object well, remove the analyzer and focus the crystals, then replace the analyzer. Cross the Nicols. In the dark field will be seen multitudes of shining crystals, and if the preparation is a fresh one in water, part of the smaller crystals will alternately flash and disappear. By observing carefully, some of the larger crystals will be found to remain dark with crossed Nicols, others will shine continuously. If the crystals are in such a position that the light passes through them parallel with the optic axis,* the crystals are isotropic like the salt crystal and remain dark. If, however, the light traverses them in any other direction the ray from the polarizer is divided into two constituents vibrating in planes at right angles to each other, and one of these will traverse the analyzer, hence such crystals will appear as if self-luminous in a dark field. The experiment with these crystals from the frog succeeds well with a 2 mm. homogeneous immersion.

As further illustration of anisotropic objects, mount some cotton fibers in balsam (Ch. VII), also some of the lens paper (§ 107). These furnish excellent examples of vegetable fibers.

Striated muscular fibers are also very well adapted for polarizing objects.

As examples of biaxial crystals, allow some borax solution to dry and crystallize on a slide; use the crystals as object. As all doubly refracting objects restore the light with crossed Nicols, they are sometimes called depolarizing.

(C) *Pleochroism.*—This is the exhibition of different tints as the analyzer is rotated. An excellent subject for this will be found in blood crystals.

(D) For the aid given by the polariscope in micro-chemistry, see (Ch. VII).

(E) See works on petrology and mineralogy for the application of the micro-polarizer in those subjects.

§ 218. **Production of Colors.**—For the production of gorgeous colors, a plate of selenite giving blue and yellow colors is placed between

---

*The optic axis of doubly refracting crystals is the axis along which the crystal is not doubly refracting, but isotropic like glass. When there is but one such axis, the crystal is said to be uniaxial, if there are two such axes the crystal is said to be bi-axial.

The crystals of carbonate of lime from the frog (see § 142) are uniaxial crystals. Borax crystals are bi-axial.

the polarizer and the object. If properly mounted, the selenite is very conveniently placed on the diaphragm carrier of the Abbe illuminator, just above the polarizer. A thin plate or film of mica also answers well.

It is not necessary to use a selenite or piece of mica for the production of the most glorious colors in many objects. One of the most beautiful preparations, and one of the most instructive also, may be prepared as follows: Heat some xylene balsam on a slide until the xylene is nearly evaporated. Add some crystals of the hypnotic medicine, sulphonal and warm till the sulphonal is melted and mixes with the balsam. While the balsam is still melted put on a cover-glass. If one gets perfect crystals there will be shown not only most beautiful colors, but the black cross with perfection. (Clark).

It is very instructive and interesting to examine organic and inorganic substances with a micro-polarizer. If the objects enumerated in § 144 were all examined with polarized light an additional means of detecting them would be found.

REFERENCES TO THE POLARISCOPE AND TO THE USE OF POLARIZED LIGHT.

Anthony & Brackett; Behrens, 133; Behrens, Kossel und Schiefferdecker; Carnoy, 61; Carpenter-Dallinger, 262, 269, 992; Clark; Daniell, 494; Davis; v. Ebener; Gage; Gamgee; Halliburton, 36, 272; Hogg, 133, 729; Lehmann; M'Kendrick; Nägeli und Schwendener, 299; Queckett; Suffolk, 125; Valentin. Physical Review, I., p. 127. Daniell, Physics for Medical Students.

CHAPTER VII.

SLIDES AND COVER-GLASSES; MOUNTING; ISOLATION, SECTIONING BY THE COLLODION AND THE PARAFFIN METHODS; LABELING AND STORING MICROSCOPICAL PREPARATIONS; EXPERIMENTS IN MICROCHEMISTRY.

APPARATUS AND MATERIAL FOR THIS CHAPTER.

Microscope, compound and simple (Ch. I); Micro-Spectroscope and polariscope (Ch. VI); Slides and cover-glasses (§ 219-220); Cleaning mixtures for glass (§ 227); Alcohol and distilled or filtered water (§ 222); fine forceps for handling cover-glasses (§ 222-226); Old handkerchiefs or lens paper (§ 107, 223). Paper boxes for storing cover-glasses (§ 223, 225); Cover-glass measurer (Figs. 120-122); Mounting material,—Farrant's solution, glycerin, glycerin-jelly and Canada balsam (§ 243, 246); Centering card and card for serial sections (§ 236); Material for dissociation and for the paraffin and collodion method (§ 244); Material for paraffin and collodion sectioning (§ 250); Net-micrometer for arranging minute objects like diatoms (§ 317); Labels (§ 309); Carbon ink for writing labels (§ 295); Writing diamond (§ 295); Shellac cement (§ 316); Cabinet (§ 296); Re-agents for experiments in micro-chemistry (§ 315).

SLIDES AND COVER-GLASSES.

§ 219. **Slides, Glass Slides or Slips, Microscopic Slides or Slips.** These are strips of clear flat glass upon which microscopic specimens are usually mounted for preservation and ready examination. The size that has been almost universally adopted for ordinary preparations is 25 x 76 millimeters (1 x 3 inches). For rock sections, slides 25 x 45 mm. or 32 x 32 mm. are used; for serial sections, slides 25 x 76 mm., 50 x 75 mm. or 37 x 87 mm. are used. For special purposes, slides of the necessary size are employed without regard to any conventional standard.

Whatever size of slide is used, it should be made of clear glass and the edges should be ground. It is altogether false economy to mount microscopic objects on slides with unground edges. It is unsafe also as the unground edges are liable to wound the hands.

§ 220. **Cleaning Slides.**—For new slides a thorough rinsing in clean water with subsequent wiping with a soft towel, and then an old soft handkerchief, usually fits them for ordinary use. If they are not satisfactorily cleaned in this way, soak them a short time in 50% or 75% alcohol, let them drain for a few moments on a clean towel or on blotting paper, and then wipe with a soft cloth. In handling the slides grasp them by their edges to avoid soiling the face of the slide. After the slides are cleaned they should be stored in a place as free as possible from dust.

For used slides, if only water, glycerin or glycerin jelly has been used on them, they may be cleaned with water, or preferably, warm water and then with alcohol if necessary. Where balsam, or any oily or gummy substance has been used upon the slides, they may be freed from the balsam, etc., by soaking them for a week or more in one of the cleaning mixtures for glass. If they are first soaked in xylene, benzin or turpentine to dissolve the balsam, then soaked in the cleaning mixture, the time required will be much shortened (§ 227). After all foreign matter is removed the slides should be very thoroughly rinsed in water to remove all the cleaning mixture. They may then be treated as directed for new slides.

If slides with large covers, as in mounted series, are put into the cleaning mixture, the swelling of the balsam is liable to break the covers. Dissolving away the balsam with turpentine, etc., avoids this, and greatly shortens the time necessary for cleaning the old slides and covers.

Another excellent method for balsam mounts is to heat the slides until the balsam is soft and then remove the cover-glasses. The cleaning mixture can then act on the entire surface. It should be said, however, that at the present price of slides and cover-glasses it is hardly worth while to clean those that have been used in balsam mounting.

§ 221. **Cover-Glasses or Covering Glasses.**—These are circular or quadrangular pieces of thin glass used for covering and protecting microscopic objects. They should be very thin, $\frac{1.0}{100}$ to $\frac{2.5}{100}$ millimeter (see table, § 27). It is better never to use a cover-glass over $\frac{2.0}{100}$ mm. thick, then the preparation may be studied with a 2 mm. oil immersion as well as with lower objectives. Except for objects wholly unsuited for high powers, it is a great mistake to use cover-glasses thicker than the working distance of a homogeneous objective (§ 47). Indeed, if one wishes to employ high powers, the thicker the sections the thinner should be the cover-glass (see § 235).

The cover-glass should always be considerably larger than the object over which it is placed.

§ 222. **Cleaning Cover-Glasses.**—New cover-glasses should be put into a glass dish of some kind containing one of the cleaning mixtures

(§ 227) and allowed to remain a day or longer. In putting them in, push one in at a time and be sure that it is entirely immersed, otherwise they adhere very closely, and the cleaning mixture is unable to act freely. Soiled covers should be left a week or more in the cleaning mixture. An indefinite sojourn in the cleaner does not seem to injure the slides or covers. After one day or longer, pour off the cleaning mixture into another glass jar, and rinse the cover-glasses, moving them around with a gentle rotary motion. Continue the rinsing until all the cleaning mixture is removed. One may rinse them occasionally, and in the meantime allow a very gentle stream of water to flow on them, or they may be allowed to stand quietly and have the water renewed from time to time. When the cleaning mixture is removed rinse the covers well with distilled water, and then cover them with 50% to 75% alcohol.

§ 223. **Wiping the Cover-Glasses.** — When ready to wipe the cover-glasses, remove several from the alcohol and put them on a soft, dry cloth, or on some of the lens paper to let them drain. Grasp a cover-glass by its edges, cover the thumb and index of the other hand with a soft, clean cloth or some of the lens paper. Grasp the cover between the thumb and index and rub the surfaces. In doing this it is necessary to keep the thumb and index well opposed on directly opposite faces of the cover so that no strain will come on it, otherwise the cover is liable to be broken.

When a cover is well wiped, hold it up and look through it toward some dark object. The cover will be seen partly by transmitted and partly by reflected light, and any cloudiness will be easily seen. If the cover does not look clear, breathe on the faces and wipe again. If it is not possible to get a cover clear in this way it should be put again into the cleaning mixture.

As the covers are wiped, put them in a clean paper box. Handle them always by their edges, or use fine forceps. Do not put the fingers on the faces of the covers, for that will surely cloud them. Wood-pulp paper, from which most of the boxes are now made, constantly sheds particles into the boxes and thus soils the covers stored in them. This can be largely obviated by coating the inside of the boxes with a thin solution of shellac.

§ 224. **Cleaning Large Cover-Glasses.** — For serial sections and especially large sections, large quadrangular covers are used. These are to be put one by one into cleaning mixture as for the smaller covers and treated in every way the same. In wiping them one may proceed as for the small covers, but special care is necessary to avoid breaking them. A safe and good way to clean the large covers is to take two perfectly flat, smooth blocks, considerably larger than the cover-glasses. These blocks are covered with soft clean cloth, or with several thicknesses of the lens paper; if now the cover-glass is placed on the one block and rubbed with the other, the cover may be cleaned as by rubbing its faces with the cloth-covered finger and thumb. It is especially desirable that these large covers should be thin—not over $\tfrac{15}{100}$ to $\tfrac{20}{100}$ mm.—otherwise high objectives cannot be used in studying the preparations.

§ 225. **Measuring the Thickness of Cover-Glasses.**—It is of the greatest advantage to know the exact thickness of the cover-glass on an object; for, (a) One would not try to use objectives in studying the preparation of a shorter working distance than the thickness of the cover (§ 57); (b) In using adjustable objectives with the collar graduated for different thicknesses of cover, the collar might be set at a favorable point without loss of time; (c) For unadjustable objectives the thickness of cover may be selected corresponding to that for which the objective was corrected (see table, § 27). Furthermore, if there is a variation from the standard, one may remedy it, in part at least, by lengthening the tube if the cover is thinner, and shortening it if the cover is thicker than the standard (§ 96).

In the so-called No. 1 cover-glasses of the dealers in microscopical supplies, the writer has found covers varying from $\tfrac{10}{100}$ mm. to $\tfrac{35}{100}$ mm. To use cover-glasses of so wide a variation in thickness without knowing whether one has a thick or thin one is simply to ignore the fundamental principles on which correct microscopic images are obtained.

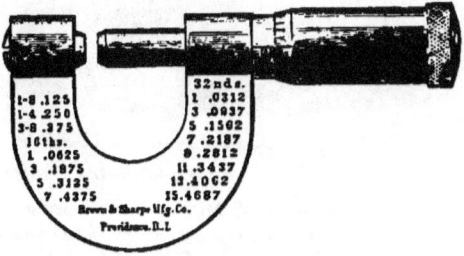

FIG. 120. *Micrometer Calipers (Brown and Sharpe). Pocket Calipers, graduated in inches or millimeters, and well adapted for measuring cover-glasses.*

It is then strongly recommended that every preparation shall be covered with a cover-glass whose thickness is known, and that this thickness should be indicated in some way on the preparation.

§ 226. **Cover-Glass Measurers.**—For the purpose of measuring cover-glasses there are three very excellent pieces of apparatus. The micrometer calipers, used chiefly in the mechanic arts, is convenient, and from its size easily carried in the pocket. The two cover-glass measurers, specially designed for the purpose, are shown in Figs. 120-122. With either of these the covers may be more rapidly measured than with the calipers.

With all of these measurers or gauges one should be certain that the index stands at zero when at rest. If the index does not stand at zero it should be adjusted to that point, otherwise the readings will not be correct.

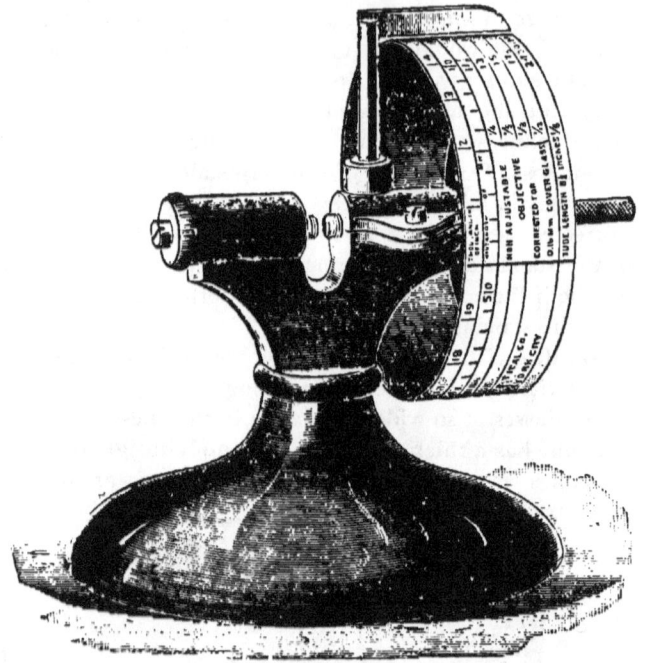

FIG. 121. *Cover Glass Measurer (Edward Bausch).*

The cover glass is placed in the notch between the two screws, and the drum is turned by the milled head at the right till the cover is in contact with the screws. The thickness is then indicated by the knife edge on the drum and may be read off directly in $\frac{1}{1000}$th mm. or $\frac{1}{1000}$th inch. In other columns is given the proper tube-length for various unadjustable objectives ($\frac{1}{2}$, $\frac{1}{4}$, $\frac{1}{5}$, and $\frac{1}{12}$ in.) made by the Bausch and Lomb Optical Company.

As the covers are measured the different thicknesses should be put into different boxes and properly labeled. Unless one is striving for the most accurate possible results, cover-glasses not varying more than

$\frac{1}{100}$ mm. may be put in the same box. For example, if one takes $\frac{1.3}{100}$ mm. as a standard, covers varying $\frac{2}{100}$ mm. on each side may be put into the same box. In this case the box would contain covers of $\frac{1.3}{100}$, $\frac{14}{100}$, $\frac{1.5}{100}$, $\frac{1.6}{100}$, and $\frac{1.7}{100}$ mm.

FIG. 122. *Zeiss Cover-Glass Measurer. With this the knife edge jaws are opened by means of a lever, and the cover inserted. The thickness may then be read off on the face as the pointer indicates the thickness in hundredths millimeter in the outer circle and in hundredths inch on the inner circle.*

§ 227. **Cleaning Mixtures for Glass**.—The cleaning mixtures used for cleaning slides and cover-glasses are those commonly used in chemical laboratories :

(A) *Dichromate of Potash and Sulphuric Acid.*
Dichromate of potash ($K_2 Cr_2 O_7$) - - - 200 grams.
Water, distilled or ordinary - - - 1000 cc.
Sulphuric acid ($H_2 SO_4$) - - - - 1000 cc.

Dissolve the dichromate in the water by the aid of heat. Pour the solution into a bottle that has been warmed and surrounded by a wet towel. Add slowly and at intervals the sulphuric acid. It is safer to mix the ingredients in an agate-ware basin, and put into the bottle only after the mixture is cool.

For making this mixture, ordinary water, commercial dichromate and strong commercial sulphuric acid should be used. It is not necessary to employ chemically pure materials.

This is a very excellent cleaning mixture, and is practically odorless. It is exceedingly corrosive and must be kept in glass vessels. It may be used more than once, but when the color changes markedly from that seen in the fresh mixture it should be thrown away.

(B) *Sulphuric and Nitric Acid Mixture.*
Nitric acid ($H NO_3$) - - - - - 200 cc.
Sulphuric acid ($H_2 SO_4$) - - - - - 300 cc.

The acids should be strong, but they need not be chemically pure. The two acids are mixed slowly, and kept in a glass-stoppered bottle. This is a more corrosive mixture than (A), and has the undesirable feature of giving off very stifling fumes, therefore it must be carefully covered. It may be used several times. It acts more rapidly than the dichromate mixture, but on account of the fumes is not so well adapted for general laboratories.

MOUNTING, AND PERMANENT PREPARATION OF MICROSCOPICAL OBJECTS.

§ 228. **Mounting a Microscopical Object** is so arranging it upon some suitable support (glass slide) and in some suitable mounting medium that it may be satisfactorily studied with the microscope.

The cover-glass on a permanent preparation should always be considerably larger than the object; and where several objects are put under one cover-glass it is false economy to crowd them too closely together.

§ 229. **Temporary Mounting.**—For the study of living objects, like amoebae, white blood corpuscles, and many other objects both animal and vegetable, their living phenomena can best be studied by mounting them in the natural medium. That is, for amoebae, in the water in which they are found; for the white blood corpuscles, a drop of blood is used and, as the blood soon coagulates, they are in the serum. Sometimes it is not easy or convenient to get the natural medium, then some liquid that has been found to serve in place of the natural medium is used. For many things, water with a little common salt (water 100 cc., common salt $\frac{6}{10}$ths gram) is employed. This is the so-called normal salt or saline solution. For the ciliated cells from frogs and other amphibia, nothing has been found so good as human spittle. Whatever is used, the object is put on the middle of the slide and a drop of the mounting medium added, and then the cover-glass. The cover is best put on with fine forceps, as shown in Fig. 123. After the cover is in place, if the preparation is to be studied for some time, it is better to avoid currents and evaporation by painting a ring of castor oil around the cover in such a way that part of the ring will be on the slide and part on the cover (Fig. 140).

FIG. 123. *To show the method of putting a cover-glass upon a microscopic preparation. The cover is grasped by one edge, the opposite edge is then brought down to the slide, and the cover gradually lowered upon the object.*

FIG. 124. *Needle Holder (Queen & Co.). By means of the screw clamp or chuck at one end, the needle may be quickly changed.*

§ 230. **Permanent Mounting.**—For making permanent microscopical preparations, there are three great methods. Special methods of procedure are necessary to mount objects successfully in each of these ways. The best mounting medium and the best method of mounting in a given case can only be determined by experiment. In most cases some previous observer has already made the necessary experiments and furnished the desired information.

The three methods are the following: (A) *Dry or in air* (§ 231); (B) *In some medium miscible with water, as glycerin or glycerin jelly* (§ 235); (C) *In some resinous medium like dammar or Canada balsam* (§ 240).

§ 231. **Mounting Dry or in Air.**—The object should be thoroughly dry. If any moisture remains it is liable to cloud the cover-glass, and the specimen may deteriorate. As the specimen must be sealed, it is necessary to prepare a cell slightly deeper than the object is thick. This is to support the cover-glass, and also to prevent the running in by capillarity of the sealing mixture.

ORDER OF PROCEDURE IN MOUNTING OBJECTS DRY OR IN AIR.

1. A cell of some kind is prepared. It should be slightly deeper than the object is thick (§ 233).
2. The object is thoroughly dried (desiccated) either in dry air or by the aid of gentle heat.
3. If practicable the object is mounted on the cover-glass; if not it is placed in the bottom of the cell.
4. The slide is warmed till the cement forming the cell wall is somewhat sticky, or a very thin coat of fresh cement is added; the cover is warmed and put on the cell and pressed down all around till a shining ring indicates its adherence (§ 234).
5. The cover-glass is sealed (§ 234).
6. The slide is labeled (§ 292).
7. The preparation is cataloged and safely stored (§ 293, 296).

§ 232. **Example of Mounting Dry, or in Air.**—Prepare a shallow cell and dry it (§ 233). Select a clean cover-glass slightly larger than the cell. Pour upon the cover a drop of a 10% solution of salicylic acid in 95% alcohol. Let it dry spontaneously. Warm the slide till the cement ring or cell is somewhat sticky, then warm the cover gently and put it on the cell, pressing down all around (§ 231). Seal the cover, label and catalog (§ 234, 292, 293).

A preparation of mammalian red blood corpuscles may be made very satisfactorily by spreading a very thin layer of fresh blood on a cover with the end of a slide. After it is dry, warm gently to remove the last traces of moisture and mount precisely as for the crystals. One can get the blood as directed for the Micro-spectroscopic work (§ 201).

FIG. 125. *Turn-Table for sealing cover-glasses and making shallow mounting cells.* (*Queen & Co.*)

§ 233. **Preparation of Mounting Cells.**—(A) *Thin Cells.* These are most conveniently made of some of the microscopical cements. Shellac is one of the best and most generally applicable (§ 316). To prepare a shellac cell place the slide on a turn-table (Fig. 125) and center it, that is, get the center of the slide over the center of the turn-table. Select a guide ring on the turn-table which is a little smaller than the cover-glass to be used, take the brush from the shellac, being sure that there is not enough cement adhering to it to drop. Whirl the turn-table and hold the brush lightly on the slide just over the guide ring selected. An even ring of the cement should result. If it is uneven, the cement is too thick or too thin, or too much was on the brush. After a ring is thus prepared remove the slide and allow the cement to dry spontaneously, or heat the slide in some way. Before the slide is used for mounting, the cement should be so dry when it is cold that it does not dent when the finger nail is applied to it.

A cell of considerable depth may be made with the shellac by adding successive layers as the previous one drys.

(B) *Deep Cells* are sometimes made by building up cement cells, but more frequently, paper, wax, glass, hard rubber, or some metal is used for the main part of the cell. Paper rings, block tin or lead rings are easily cut out with gun punches. These rings are fastened to the slide by using some cement like the shellac.

§ 234. **Sealing the Cover-Glass for Dry Objects Mounted in Cells.**—When an object is mounted in a cell, the slide is warmed until the cement is slightly sticky, or a very thin coat of fresh cement is put on. The cover-glass is warmed slightly also, both to make it stick to the cell more easily, and to expel any remaining moisture from the object. When the cover is put on it is pressed down all around over the cell until a shining ring appears, showing that there is an intimate contact. In doing this use the convex part of the fine forceps or some other blunt, smooth object; it is also necessary to avoid pressing on the cover except immediately over the wall of the cell for fear of breaking the cover. When the cover is in contact with the wall of cement all around, the slide should be placed on the turn-table and carefully arranged so that the cover-glass and cell wall will be concentric with the guide rings of the turn-table. Then the turn-table is whirled and a ring of fresh cement is painted, half on the cover and half on the cell wall (Fig. 140). If the cover-glass is not in contact with the cell wall at any point and the cell is shallow, there will be great danger of the fresh cement running into the cell and injuring or spoiling the preparation.

When the cover-glass is properly sealed, the preparation is put in a safe place for the drying of the cement. It is advisable to add a fresh coat of cement occasionally.

§ 235. **Mounting Objects in Media Miscible with Water.**— Many objects are so greatly modified by drying that they must be mounted in some medium other than air. In some cases water with something in solution is used. Glycerin of various strengths, and glycerin jelly are also much employed. All these media keep the object moist and therefore in a condition resembling the natural one. The object is usually and properly treated with gradually increasing strengths of glycerin or fixed by some fixing agent before being permanently mounted in strong glycerin or either of the other media.

In all of these different methods, unless glycerin of increasing strengths has been used to prepare the tissue, the fixing agent is washed away with water before the object is finally and permanently mounted in either of the media.

For glycerin jelly no cell is necessary unless the object has a considerable thickness.

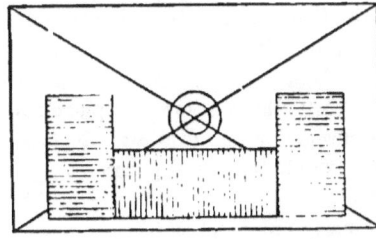

FIG. 126. *Centering Card. A card with stops for the slide and circles in the position occupied by the center of the slide. If the slide is put upon such a card it is very easy to arrange the object so that it will be approximately in the center of the slide.* (From the Microscope, Dec., 1886.)

§ 236. **Order of Procedure in Mounting Objects in Glycerin.**

1. A cell must be prepared on the slide if the object is of considerable thickness (§ 233, 234).

2. A suitably prepared object (§ 235) is placed on the center of a clean slide, and if no cell is required a centering card is employed to facilitate the centering (Fig. 126.)

3. A drop of pure glycerin is put upon the object, or if a cell is used, enough to fill the cell.

4. In putting on the cover-glass it is grasped with fine forceps and the under side breathed on to slightly moisten it so that the glycerin will adhere, then one edge of the cover is put on the cell or slide and the cover gradually lowered upon the object (Fig. 123). The cover is then gently pressed down. If a cell is used, a fresh coat of cement is added before mounting (§ 234.)

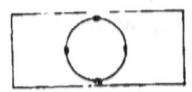

FIG. 127. *Slide and cover glass showing method of anchoring a cover-glass with a glycerin preparation when no cell is used. A cover-glass so anchored is not liable to move when the cover is being sealed* (§ 238).

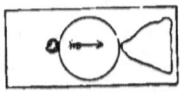

FIG. 128. *Glass slide with cover-glass, a drop of reagent and a bit of absorbent paper to show method of irrigation* (§ 247, 248).

5. The cover-glass is sealed (§ 234).
6. The slide is labeled (§ 292).
7. The preparation is cataloged and safely stored (§ 293, 296).

§ 237. **Order of Procedure in Mounting Objects in Glycerin Jelly.**

1. Unless the object is quite thick no cell is necessary with glycerin jelly.

2. A slide is gently warmed and placed on the centering card (Fig. 126) and a drop of warmed glycerin jelly is put on its center. The suitably prepared object is then arranged in the center of the slide.

3. A drop of the warm glycerin jelly is then put on the object, or if a cell is used it is filled with the medium.

4. The cover-glass is grasped with fine forceps, the lower side breathed on and then gradually lowered upon the object (Fig. 123), and gently pressed down.

5. After mounting, the preparation is left flat in some cool place till the glycerin jelly sets, then the superfluous amount is scraped and wiped away and the cover-glass sealed with shellac (§ 234, 248).

6. The slide is labeled (§ 292).
7. The preparation is cataloged and safely stored (§ 296).

§ 238. **Sealing the Cover-Glass when no Cell is used.**—(A) *For glycerin mounted specimens.* The superfluous glycerin is wiped away as carefully as possible with a moist cloth, then four minute drops of cement are placed at the edge of the cover (Fig. 127), and allowed to harden for half an hour or more. These will anchor the cover-glass, then the preparation may be put on the turn-table and a ring of cement put around the edge while whirling the turn-table.

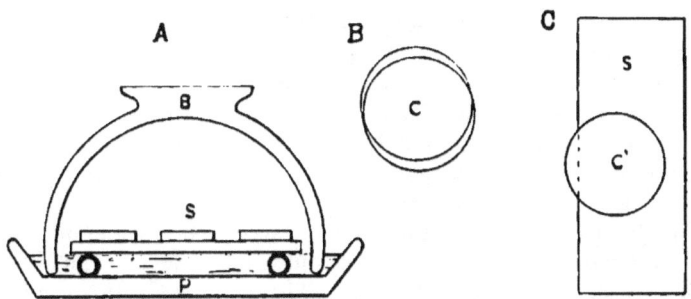

FIG. 129. A—*Simple form of moist chamber made with a plate and bowl.* B, *bowl serving as a bell jar;* P, *plate containing the water and over which the bowl is inverted;* S, *slides on which are mounted preparations which are to be kept moist. These slides are seen endwise and rest upon a bench made by cementing short pieces of large glass tubing to a strip of glass of the desired length and width.*
B—*Two cover-glasses* (C) *made eccentric, so that they may be more easily separated by grasping the projecting edge.*
C—*Slide* (S) *with projecting cover-glass* (C). *The projection of the cover enables one to grasp and raise it without danger of moving it on the slide and thus folding the substance under the cover.* (From Proc. Amer. Micr. Soc., 1891).

(B) *For objects in glycerin jelly, Farrant's solution or a resinous medium.* The mounting medium is first allowed to harden, then the superfluous medium is scraped away as much as possible with a knife, and then removed with a cloth moistened with water for the glycerin jelly and Farrant's solution or with alcohol, chloroform or turpentine, etc., if a resinous medium is used. Then the slide is put on a turn-table and a ring of the shellac cement added. (C) *Balsam preparations* may be sealed with shellac as soon as they are prepared, but it is better to allow them to dry for a few days. One should never use a cement for sealing preparations in balsam or other resinous media unless the solvent of the cement is not a solvent of the balsam, etc. Otherwise the cement will soften the balsam and finally run in and mix with it, and partly or wholly ruin the preparation. Shellac is an excellent cement for sealing balsam preparations, as it never runs in, and it serves to avoid any injury to the preparation when cedar oil, etc., are used for homogeneous immersion objectives.

§ 239. **Example of Mounting in Glycerin Jelly.**—For this select some stained and isolated muscular fibres or other suitably prepared objects. (See under isolation § 244). Arrange them on the middle of a slide, using the centering card, and mount in glycerin jelly as directed in § 223. Air bubbles are not easily removed from glycerin jelly preparations, so care should be taken to avoid them.

§ 240. **Mounting Objects in Resinous Media.**—While the media miscible with water offer many advantages for mounting animal and vegetable tissues the preparations so made are liable to deteriorate. In many cases, also, they do not produce sufficient transparency to enable one to use high enough powers for the demonstration of minute details.

By using sufficient care almost any tissue may be mounted in a resinous medium and retain all its details of structure.

For the successful mounting of an object in a resinous medium it must in some way be deprived of all water and all liquids not miscible with the resinous mounting medium. There are two methods of bringing this about: (A) By drying or desiccation (§ 241), and (B) by successive displacements (§ 243).

§ 241. **Order of Procedure in Mounting Objects in Resinous Media by Desiccation :**

1. The object suitable for the purpose (fly's wings, etc.) is thoroughly dried in dry air or by gentle heat.
2. The object is arranged as desired in the center of a clean slide on the centering card (Fig. 126).
3. A drop of the mounting medium is put directly upon the object or spread on a cover-glass.
4. The cover-glass is put on the specimen with fine forceps (Fig. 123), but in no case does one breathe on the cover as when media miscible with water are used.
5. The cover-glass is pressed down gently.
6. The slide is labeled (§ 292).
7. The preparation is cataloged and safely stored (§ 293, 296).
8. Although it is not absolutely necessary, it is better to seal the cover with shellac after the medium has hardened round the edge of the cover (§ 238 C).

§ 242. **Example of Mounting in Balsam by Desiccation.**—Find a fresh fly, or if in winter, procure a dead one from a window sill or a spider's web. Carefully remove the fly's wings, being especially careful to keep them the dorsal side up. With a camel's hair brush remove any dirt that may be clinging to them. Place a clean slide on the centering card, then with fine forceps put the two wings within one of the guide rings. Leave one dorsal side up, turn the other ventral side up. Spread some Canada balsam on the face of the cover glass and with the fine forceps place the cover upon the wings (Fig. 123). Probably some air-bubbles will appear in the preparation, but if the slide is put in a warm place these will soon disappear. Label, catalog, etc., (§ 291-295).

§ 243. **Mounting in Resinous Media by a Series of Displacements.**—For examples of this see the procedure in the paraffin and in the collodion methods (§ 265, 284). The first step in the series is *Dehydration*, that is, the water is displaced by some liquid which is miscible both with the water and the next liquid to be used. Strong alcohol (95% or stronger) is usually employed for this. Plenty of it must be used to displace the last trace of water. The tissue may be soaked in a dish of the alcohol, or alcohol from a pipette may be poured upon it. Dehydration usually occurs in the thin objects to be mounted in balsam in 5 to 15 minutes. If a dish of alcohol is used it must not be used too many times, as it loses in strength.

The second step is clearing. That is, some liquid which is miscible with the alcohol and also with the resinous medium is used. This liquid is highly refractive in most cases, and consequently this step is called *clearing* and the liquid a *clearer*. The clearer displaces the alcohol, and renders the object more or less translucent. In case the water was not all removed, a cloudiness will appear in parts or over the whole of the preparation. In this case the preparation must be returned to alcohol to complete the dehydration.

One can tell when a specimen is properly cleared by holding it over some dark object. If it is cleared it can be seen only with difficulty, as but little light is reflected from it. If it is held toward the window, however, it will appear translucent.

The third and final step is the displacement of the clearer by the resinous mounting medium.

The specimen is drained of clearer and allowed to stand for a short time till there appears the first sign of dullness from evaporation of the clearer from the surface. Then a drop of the resinous medium is put on the object, and finally a cover-glass is placed over it, or a drop of the mounting medium is spread on the cover and it is then put on the object.

ISOLATION OF HISTOLOGICAL ELEMENTS.

§ 244. For a correct conception of the forms of the cells and fibers of the various organs of the body, one must see these elements isolated and thus be able to inspect them from all sides. It frequently occurs also that the isolation is not quite complete, and one can see in the clearest manner the relations of the cells or fibers to one another.

The chemical agents or solutions for isolating are, in general, the same as those used for hardening and fixing. But the solutions are only about one-tenth as strong as for fixing, and the action is very much shorter, that is, from one or two hours to as many days. In the weak solution the cell cement or connective tissue is softened so that the cells and fibers may be separated from one another, and at the same time the cells are preserved. In fixing and hardening, on the other hand, the cell cement, like the other parts of the tissue, are made firmer. It is better also to dilute the fixing agents with normal salt solution (§ 313) than merely with water.

§ 245. **Isolation by Means of Formaldehyde.**—Formaldehyde in a $\frac{1}{20}\%$ solution in normal salt solution is one of the very best dissociating agents for brain tissue and all the forms of epithelium (§ 308). It is prepared as follows: 5 cc. of formal, formol, formalin or formalose, that is, a 40% solution of formaldehyde, are mixed with 995 cc. of normal salt solution. This acts quickly and preserves delicate structures like the cilia of ordinary epithelia, and also of the endymal cells of the brain. It is very satisfactory for isolating the nerve cells of the brain. For the epithelium of the trachea, intestines, etc., the action is sufficient in two hours; good preparations may also be obtained after two days or more. The action on nerve tissue of the brain is about as rapid. For the stratified epithelia, like those of the skin, mouth, etc., it may require two or three days for the most satisfactory preparations. See Figs. 130 and 131.

§ 246. **Example of Isolation.**—Place a piece of the trachea of a very recently killed animal, or the roof of a frog's mouth, in the formaldehyde dissociator. After two hours or more, up to two or three days, excellent preparations of ciliated cells may be obtained by scraping the trachea or roof of the mouth and mounting the scrapings on a slide. If one proceeds after two hours, probably most of the cells will cling together, and in the various clumps will appear cells on end showing the cilia or the bases of the cells, and other clumps will show the cells in profile. By tapping the cover gently with a needle holder or other light object the cells will be more separated from one another, and many fully isolated cells will be seen.

§ 247. **Staining the Cells.**—Almost any stain may be used for the formalin dissociated cells. As an example, one may use eosin (§ 305). This may be drawn under the cover of the already mounted preparation (Fig. 128), or a new preparation may be made and the scrapings mixed with a drop of the eosin before putting on the cover-glass. It is an advantage to study unstained preparations, otherwise one may obtain the erroneous opinion that the structure cannot be seen unless it is stained. The stain makes the structural features somewhat plainer; it also accentuates some features and does not affect so markedly others.

§ 248. **Permanent Preparations of Isolated Cells.**—If one desires to make a permanent preparation of the isolated cells it may be done by placing a drop of glycerin at the edge of the cover and allowing it to diffuse under the cover, or the diffusion may be hurried by using

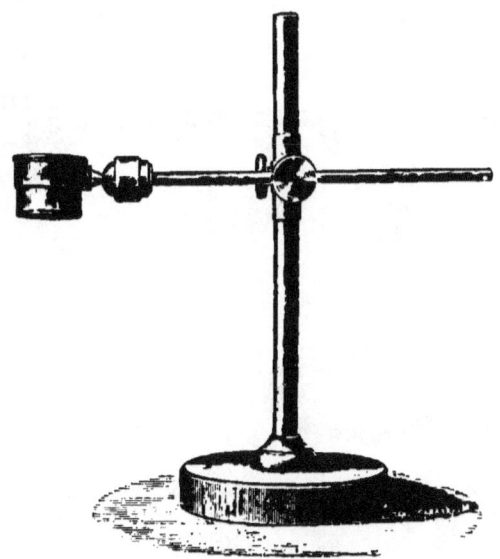

FIG. 130. *Adjustable lens holder with universal joint. This is especially useful for gross dissections, and for dissecting the partly isolated elements with needles.*

a piece of blotting paper, as shown in Fig. 128. One may also make a new preparation and either with or without staining, mix the cells with a drop of glycerin on the slide and then cover, or one may use glycerin jelly (§ 239, 309).

§ 249. **Isolation of Muscular Fibers.**—For this the formal dissociator may be used (§ 245, 308), but the nitric acid method is more successful § 312   The fresh muscle is placed in this in a glass vessel. At the ordinary temperature of a sitting room (20 degrees centigrade) the connective tissue will be so far gelatinized in from one to three days that it is very easy to separate the fascicles and fibers either with needles or by shaking in a test tube or reagent vial (Fig. 132) with water. It takes longer for some muscles to dissociate than others, even in the same temperature, so one must try occasionally to see if the action is sufficient. When it is, the acid is poured off and the muscle washed

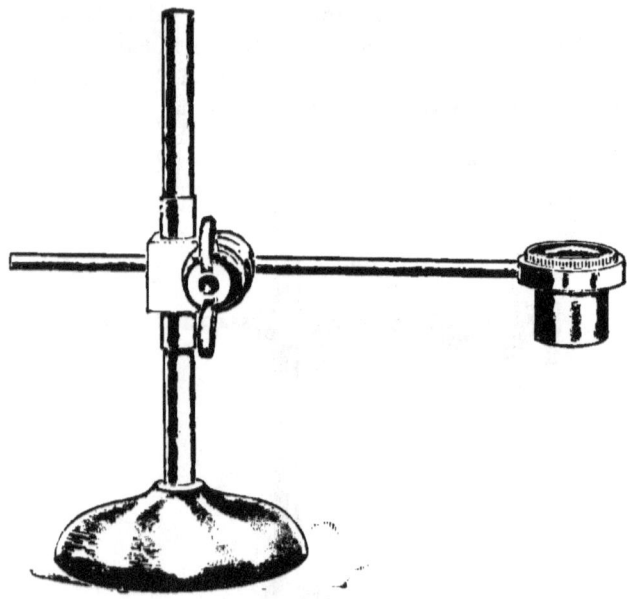

FIG. 131. *Adjustable lens holder for the same purposes as Fig. 130.* (*The Bausch & Lomb Optical Company*).

gently with water to remove the acid. If one is ready to make the preparations at once they may be isolated and mounted in water. If it is desired to keep the specimen indefinitely, or several days, the water should be poured off and a half saturated solution of alum added (§ 299). The alum solution is also very advantageous if the specimens are to be stained. The specimens may be mounted in glycerin, glycerin jelly or balsam. Glycerin jelly is the most satisfactory, however.

THE PREPARATION OF SECTIONS OF TISSUES AND ORGANS.

§ 250. At the present time there are three principal methods of obtaining thin sections of tissues and organs for microscopic study. These methods are : *The Collodion Method, the Paraffin Method, and the Freezing Method.* Each of these methods has its special application, although the collodion method is perhaps the most generally applicable, and the freezing method the most restricted, and is used mostly in pathological work, where rapid diagnosis is necessary and the finest details of structure are not so important. With the paraffin method the thinnest sections may be made, and in some ways it is the most satisfactory of all. A good microtome is of very great aid in sectioning.

§ 251. **The Collodion Method.**—In sectioning by this method the tissues are first hardened properly and then entirely infiltrated with collodion, and the collodion hardened. It is not removed from the tissue, but on account of its transparency does no harm.

§ 252. **Fixing and Hardening the Tissue.**—Any of the approved methods of hardening and fixing may be employed. A good general method which is applicable to nearly all of the tissues and organs is that by Picric-Alcohol. For the preparation of the solution see (§ 315). A small piece of tissue or organ not containing more than two to three cubic centimeters is placed in 40 or 50 cc. of the picric-alcohol and left 6 to 24 hours, when the first picric-alcohol should be thrown away and fresh added. After one or two days more the picric-alcohol should be poured off and 67% alcohol added. In a day or two this is replaced by 75% or 82% alcohol ; 82% is on the whole most satisfactory, and the tissue may be left in this till it is ready for dehydration.

§ 253. **Dehydration before Infiltration.**—When one is ready to imbed for sections, the tissue must first be dehydrated in plentiful 95% or stronger alcohol. It is better to take only a small piece for this. The smaller the piece the thinner the sections may be made. The dehydration will usually be completed in 2 to 24 hours. If the alcohol is changed two or three times the dehydration will be hastened.

§ 254. **Saturating with Ether-Alcohol** (§ 306).—The next step is to remove the tissue from the alcohol and place it in a vial of ether-alcohol (§ 306) for 2 to 24 hours. The dehydration is somewhat more complete by this step, and the tissue is more perfectly prepared for the reception of the collodion. If the dehydration is very thorough in the alcohol, this step may be omitted, however, but one is surer of success if the ether-alcohol is used.

§ 255. **Infiltration with Thin Collodion.**—The ether-alcohol is poured off, and a mixture of thin collodion is added (§ 304). Two or three hours will suffice for objects two or three millimeters in thickness. A stay of one or more days does no harm. The larger the object the more time is needed.

§ 256. **Infiltration with Thick Collodion.**—The thin collodion is poured off and thick collodion (§ 304) added. For very small objects, four or five hours will suffice to infiltrate, but for larger objects a longer time is necessary. The tissue does not seem to be injured at all in the thick collodion, and a stay in it during a day or even a week is more certain to insure a perfect infiltration.

§ 257. **Imbedding.**—The tissue may be imbedded in a paper box, such as is used for paraffin imbedding, or in any of the other boxes devised for paraffin. It is better, if paper is used, to put a very small amount of oil on the paper to prevent the collodion from sticking to it. Vaselin spread over lightly and then all removed, so far as possible, with a cloth or with lens paper, gives the right surface. For small objects it is more convenient to imbed immediately on a holder that may be clamped into the microtome. Cylinders or blocks of glass, vulcanite, wood and cork have all been recommended and used. A cork of the proper size is most convenient, and for many purposes answers well. Some collodion is put on the end of the cork and a pin put near one edge. The tissue is transferred from the thick collodion to the cork and leaned against the pin. Drops of the thick collodion are then poured on the tissue, and by moving the cork properly the thick, viscid mass may be made to surround and envelop the tissue. Drops of collodion are added at short intervals until the tissue is well surrounded, and then as soon as a slight film hardens on the surface, the cork bearing the tissue is inverted in a wide-mouth vial of considerably larger diameter than the cork (Fig. 132). The vial should contain sufficient chloroform to float the cork. The vial is then tightly corked. In imbedding somewhat larger objects on the end of a cork or other holder, it is frequently advantageous to wind oiled paper around the holder or cork, tie it tightly and have the projecting hollow cylinder sufficiently long to receive the object. The tissue is then put into the cylinder and sufficient collodion added to completely immerse it. As soon as a film has formed over the exposed end, the cork may be inverted and immersed in chloroform, as described above.

§ 258. **Hardening and Clarifying the Collodion.**—After a few hours the collodion is hardened by the chloroform. If it acts long enough the imbedding mass is rendered entirely transparent, if no water is present. Whenever the collodion is hard, whether it is clear or not, the chloroform is poured off and the carbol-xylene* clarifier (§ 302 added. In a few hours the imbedded mass will become as transparent as glass and the tissue will seem to have nothing around it. Sometimes the collodion remains white and opaque for a considerable time. So far as the writer has been able to judge, this is due to moisture. If one breathes on the mass too much while imbedding, or if it is very damp in the room, the opacity may result. Sometimes, in objects of considerable size, this may remain for a week. This is the exception, however, and if the mass seems sufficiently hard and tough, the cutting may proceed even if the clarification is incomplete.†

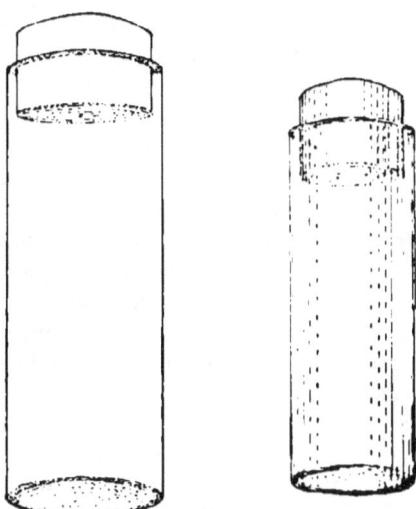

FIG. 132. *Preparation Vials for Histology and Embryology. These represent the two vials, natural size, that have been found most useful. They are kept in blocks with holes of the proper size.*

In case the imbedding mass will not clarify after a few days the imbedded object may be placed in 95% alcohol for a day for dehydration, and then passed through chloroform and into the clarifier. There is usually no trouble in getting the mass perfectly clear in this way.

*The hydrocarbon xylene (C_8H_{10}) is called xylol in German. In English, members of the hydrocarbon series have the termination "ene," while members of the alcohol series terminate in "ol."

†The imbedded object may remain in the castor-xylene clarifier indefinitely without harm. The collodion grows somewhat tougher by a prolonged stay in it. After cutting all the sections desired at one time, the imbedded tissue is returned to the clarifier for future sections.

§ 259. **Cutting the Sections.**—For collodion sectioning a long, drawing cut is necessary in order to obtain thin, perfect sections. The object is, therefore, put in the jaws of the microtome at the right level, and the knife arranged so that half or more of the blade of the knife is used in cutting the section. It is advantageous also to have the object placed with its long diameter parallel with the edge of the knife. The surrounding collodion mass should be cut away, as in sharpening a lead pencil, so that there is not more than a thickness of about two millimeters all around the tissue. This is to render the diameter of the end to be cut as small as possible. The smaller the object the thinner can the sections be made. With an object two to three millimeters thick and not over five millimeters wide, and a good sharp knife, sections 5µ to 6µ can be cut without difficulty. When knife and tissue are properly arranged the tissue is well wet and the knife flooded with the clarifier. Make the sections with a steady motion of the knife. Then draw the section up toward the back of the knife with an artist's brush and make the next section. Arrange the sections in serial order on the knife-blade till enough are cut to fill the area that the cover-glass will cover.

§ 260. **Transferring the Sections to the Slide.**—If the clarifier has evaporated so as to leave the sections somewhat dry on the knife, add a small amount. Take a piece of thin absorbent, close-meshed paper* about twice the size of a slide and place it directly upon the sections. Press the paper down evenly all around and then pull the paper off the edge of the knife.† The sections will adhere to the paper. Place the paper, sections down, on a slide, taking care that the sections are in the desired position on the slide. Use some ordinary lens-paper or any absorbent paper, and press it down gently upon the transfer paper. This will absorb the oil, and then the transfer paper may be lifted, with a rolling motion, from the slide. The sections will remain on the slide.

§ 261. **Fastening the Sections to the Slide.**—Drop just enough

---

* Various forms of paper have been used to handle the collodion sections. It should be moderately strong, fine meshed and not liable to shed lint, and fairly absorbent. One of the first and most successful papers recommended is "closet or toilet paper." Cigarette paper is also excellent. In my own work the silky Japanese paper, called "Usago" paper, has been found almost perfect for the purpose. Ordinary lens paper or thin blotting paper for absorbing the oil is used with it.

† If one is a long time cutting a series of sections, it sometimes occurs that the xylene evaporates, and while the sections may not look dry, they are practically in castor oil and not easily transferable. In such a case fresh clarifier or even a little xylene to thin the oil on the sections may be used. If the oil is too thick it is viscid and there is difficulty in handling the sections with the paper as they stick rather firmly to the knife.

ether-alcohol (equal parts of sulphuric ether and 95% alcohol) on the sections to moisten them. This will melt the collodion and fasten the sections to the slide. Allow the slide to remain in the air till the surface begins to look slightly dull or glazed.

Sometimes, especially when the air is moist, the sections wrinkle badly when the ether-alcohol is put on to fasten them to the slide. The excessive wrinkling can be avoided by using one part alcohol and two parts ether instead of using equal parts of each. Perhaps also it would be advantageous in this case to use absolute alcohol.

FIG. 133. *Pipette for adding liquids dropwise and for washing preparations.* (*Whitall, Tatum & Co.*)

§ 262. **Removing the Oil from the Sections.**—As soon as the ether-alcohol has evaporated sufficiently to leave the surface dull, place the slide in a jar of ordinary commercial benzin. It may be left here a day or more without injury to the sections, but if moved around in the jar the oil will be removed in three to five minutes. From the benzin transfer to a jar of 95% alcohol to wash away the benzin. One may use alcohol in the beginning, but it dissolves the oil far less rapidly than the benzin. The slide may remain in the alcohol half a day or more if one wishes, but a stay of five minutes or a thorough rinsing of half a minute or so by moving the slide around in the alcohol will suffice.

§ 263. **Staining the Sections with an Alcoholic Stain.**—If an alcoholic stain containing 50% or more alcohol (for example, hydrochloric acid carmine in 70% alcohol) is used, the slide may be removed from the 95% alcohol, drained somewhat and then the stain poured upon the sections, or preferably, the slide immersed in a jar of the stain. The stain is finally washed away with 67% or stronger alcohol, the sections dehydrated in 95% alcohol, cleared and mounted in balsam.

§ 264. **Staining the Sections with an Aqueous Dye.**—In staining with a watery stain, the slide bearing the sections is transferred from the 95% alcohol and plunged into a jar of water, and either allowed to remain a few minutes or moved around in the water a moment. Then it is placed horizontally, and some of the stain placed on the sections with a pipette, or preferably, it is immersed in a jar of the stain; in case of immersion, however, the slide should stand vertically or nearly so, then any particles of dust, etc., in the stain will settle to the bottom of the vessel and not settle on the sections. When the sections are stained, usually within five minutes, they are thoroughly washed with water either by the use of a pipette or preferably by immersing in a jar of water. They may then be counterstained for half a minute with some general dye, like eosin or picric acid, or mounted with but the one stain.*

FIG. 135.

FIG. 136.

FIG. 134. *Waste Bowl with rack for supporting slides and a small funnel in which the slides stand while draining. This outfit is easily made by any tin smith. The rack is composed of two brass rods about 3 mm. in diameter. The bent end pieces are sheet brass. The funnel is made of tin, copper or brass. Either copper or brass is preferable to tin. A glass dish like that shown in Fig. 135 is better than a bowl, as it can be more readily and thoroughly cleaned. (Cut loaned by Wm. Wood & Co.).*

FIG. 135. *Round glass aquarium. This glass vessel is better than the bowl for all the uses described for the bowl.* (*Whitall, Tatum & Co.*)

FIG. 136. *Glass box with cover. These boxes may be had of various sizes and can be used advantageously for water, and for cleaning mixture for slides and cover glasses (§ 227).* (*Whitall, Tatum & Co.*)

---

* In the past the plan for changing sections from 95% alcohol to water, for example, has been to run them down gradually, using 75, 50 and 35% alcohol successively. Each percentage may vary, but the principle of a gradual passing from strong alcohol to water was advocated. On the other hand, I have found that the safest method is to plunge the slide directly into water from the 95% alcohol. The diffusion currents are almost or quite avoided in this way. There is no time for the alcohol and water to mix, the alcohol is washed away almost instantly by the flood of water. So in dehydrating after the use of watery stains, the slide is plunged quickly into a jar of 95% alcohol. The diffusion currents are avoided in the same way, for the water is removed by the flood of alcohol. This plan has been submitted to the severe test of laboratory work, and has proved itself perfectly satisfactory.

ORDER OF PROCEDURE IN MAKING MICROSCOPICAL PREPARATIONS BY THE COLLODION METHOD.

§ 265. It will be seen from this table, and sections 252-266, that it requires about five days to get a microscopical preparation if one commences with the fresh tissue. Other methods of hardening might require as many months. It is evident, therefore, that one must exercise foresight in histology or much time will be wasted.

1. Fixing and hardening the tissues (§ 252), 4 days or more.
2. Dehydrating the object to be cut in 95% or stronger alcohol (§ 253), 2-24 hours.
3. Saturating the tissue in ether-alcohol (§ 254), 2-24 hours.
4. Infiltrating with thin collodion (§ 255), 2 hours to 2 days.
5. Infiltrating in thick collodion (§ 256), 5 hours to several days.
6. Imbedding the tissue (§ 257), 15 to 20 minutes.
7. Hardening the collodion with chloroform (§ 258), 5-24 hours.
8. Clarifying and further hardening the collodion with castor-xylene (§ 258), 10-36 hours.
9. Cutting the sections (§ 259), 10 minutes to 2 hours.
10. Transferring the sections to a slide with paper (§ 260), 1 minute.
11. Fastening the sections to the slide with ether-alcohol (§ 261), 1 or 2 minutes.
12. Removing the oil from the sections with benzin and alcohol (§ 262), 3-5 minutes, or 24 hours.
13. Staining the sections with an alcoholic dye (§ 263-264), 2 minutes to 24 hours.
14. Staining the sections with an aqueous dye (§ 264), 2-10 minutes.
15. Removing the superfluous dye by washing in water or alcohol (§ 263-264), 2-5 minutes.
16. Staining with a general dye (§ 264), 15-30 seconds.
17. Washing with water or alcohol (§ 263-264), 1 to 2 minutes.
18. Dehydrating the sections in 95% alcohol (§ 266), 5 min. to 24 hours.
19. Clearing the sections (§ 266), 5 min. to 24 hours.
20. Draining the sections, 1-2 minutes.
21. Mounting in Canada balsam (§ 266), 1-2 minutes.
22. Sealing the cover-glass (§ 238), 2 minutes.
23. Labeling the preparation (§ 291), 2 minutes.
24. Cataloging the preparation (§ 294), 5-10 minutes.

§ 266. **Mounting in Balsam.**—After the sections are stained they must be dehydrated and cleared before mounting in balsam. For the dehydration the slide is plunged into a jar of 95% alcohol. For clearing after the dehydration the slide is drained of alcohol and put down flat and the clearer poured on, or the whole slide is immersed in a jar of clearer (§ 303). Clearing usually is sufficient in a few minutes; a stay of an hour or even over night does not injure most sections.

In mounting in balsam the clearer is drained away by standing the slide nearly vertically on some blotting paper, or by using the waste bowl and standing it up in the little funnel (Fig. 134). Then the balsam is put on the sections or spread on the cover-glass and that placed over the sections.

For cataloging and labeling, see §§ 291-295.

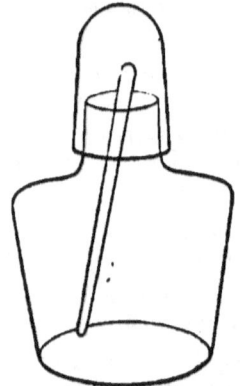

FIG. 137. *Small spirit lamp modified into a balsam bottle, or a glycerin or glycerin-jelly bottle, or a bottle for homogeneous immersion liquid. For all of these purposes it should contain a glass rod as shown in the figure. By adding a small brush, it answers well for a shellac bottle also.*

§ 267. **The Collodion Method with Alcohol.**—A good method of procedure for making collodion sections is to proceed exactly as described including § 257, and then instead of hardening the collodion in chloroform and clarifier, it is hardened in 82% alcohol for a day or two before sectioning. In sectioning, the knife and tissue are kept wet with 82% alcohol and the sections are dehydrated with 95% alcohol and then fastened to the slide with ether alone or with ether-alcohol. The staining and mounting (§ 263-266) are as described. One may preserve the tissue after imbedding for a long time in the 82% alcohol before sectioning, or for successive sections. While this method appears somewhat simpler, the results are not so satisfactory as by the oil-method given above.

THE PARAFFIN METHOD.

§ 268. As with the collodion method, the tissues are first properly fixed and hardened and then entirely filled with the imbedding mass, but unlike collodion the mass must be entirely removed before the sections are finally mounted. The tissue thus imbedded and infiltrated is like a homogeneous mass and sections may be cut of extreme thinness.

§ 269. **Harden perfectly fresh tissue** in picric-alcohol (§ 315) from one to three days. (Any good method for fixing and hardening the elements may be used. One must observe in each case, however, the special conditions necessary for each method. The time might be longer or shorter than for the picric-alcohol. See Lee, the Microtomists' Vade-Mecum.)

If picric-alcohol is used, pour it off after the proper time for fixing has elapsed, and add 67% alcohol. Leave this on the tissue from one to three days, and if it becomes very yellow it is well to change it two or three times. After two or three days pour off the 67% alcohol and add 82%. The tissue should remain in this one or two days, and it may remain indefinitely.

In case the alcohol becomes much yellowed, it should be changed.

§ 270. **Dehydration and Preparation for Imbedding**—From the pieces of tissue fixed and hardened in any approved manner, cut pieces 5 to 10 millimeters long and 2 to 3 millimeters in breadth. Place one or two pieces in a shell vial (Fig. 132) and add 95% alcohol. Change the alcohol after two or three hours, and within 6 to 24 hours, depending on the size of the piece to be dehydrated, the dehydration will be completed. The secret of success is the use of plenty of alcohol and sufficient time. Absolute alcohol for the second change would act more promptly and efficiently, but if plenty of 95% is used one will succeed, unless the day, or the climate in general, is too damp.

(If one is studying organs, then the whole organ may need to be prepared for imbedding, but for the minute structure small pieces are preferable, as thinner sections may be made.)

§ 271. **Displacing the Alcohol and Clearing Tissues with Thickened Cedar-wood Oil and Infiltrating with Paraffin.**—(Lee, p. 66. Neelson and Schiefferdecker, Arch. für Anat. und Physiol., 1882, p. 206.) When the tissue is dehydrated it is removed to a vial of thickened cedar-wood oil. When the alcohol used for dehydration is displaced by the oil, the tissue will look clear and translucent. This requires 2 to 24 hours. It is hastened by warmth. It is then removed from the cedar-wood oil, drained, and placed in pure, melted paraffin, and this is then put into a paraffin oven and left from 2 to 24 hours. It is then imbedded for sectioning.

Paraffin for infiltrating has usually a somewhat lower melting point than that for imbedding. Equal parts of paraffin of 43 C. and 54 C., answer well. For imbedding, the paraffin must be of a melting point which will give good ribbons in the temperature of the room where the sectioning is to be done. In a room of 19 to 20 C. a mixture of 1 part 43 C. paraffin with two parts of 54 C. usually answers well.

§ 272. **Imbedding in Paraffin.**—Make a small paper box, fill it about half full of pure melted paraffin of the proper melting point (see § 271), and then remove the tissue from the infiltrating oven, place it in one end of the paper box and arrange it so that sections may be made in any desired direction. As soon as the paraffin has solidified on the surface, place the box in cold water, on ice or in snow to cool the paraffin quickly, and thus avoid spaces around the tissue, etc.

§ 273. **Cutting the Sections.**—After the imbedding mass is well cooled, remove the paper box and trim the end containing the tissue in a pyramidal form. Clamp the block of paraffin in the holder of the microtome so that the tissue will be at the proper level for cutting. If a ribbon microtome is used, heat the holder and melt the end of the block upon it. Cool and place the holder in its place in the microtome. Use a very sharp, dry razor for cutting the sections. The sections are made with a rapid, straight cut as in planing. Do not try to section with a drawing cut as in collodion sectioning. If the temperature of the room is right for the paraffin used, the sections will remain flat, and if the opposite sides of the block are parallel, and one edge strikes the knife squarely, the sections will adhere and thus make a ribbon. If the room is too cold for the paraffin the sections will roll. If it is too warm the sections will be imperfect.

Remember the sections must be very thin, from 3μ to 15μ to show fine structural details to good advantage.

§ 274. **Extending the Sections with Warm Water.**—Paraffin sections are liable to be very finely wrinkled. These wrinkles in the sections often obscure the structure. To remove them, the ribbons or separate sections are placed on cold water in a dish like a waste jar (Fig. 135). Then hot water is slowly added till the sections extend. This removes the folds. When the water has cooled, the ribbons are cut into proper lengths with scissors, and the pieces transferred to albumenized slides.

§ 274a. **Extending Sections on the Slide.**—Instead of placing the sections on water in a dish, the sections may be put directly upon slides. To extend them, add sufficient water so that they will float. Warm the slide carefully until they straighten, pour off the water, and allow the slides to stand for several hours until all the water has evaporated. The sections adhere firmly to the slide and are in optical contact with it, as shown by the shiny appearance when all the water has evaporated.

§ 275. **Fastening the Sections to the Slide.**—To fasten the sections firmly to the slide, coat the slide with albumen fixative (§ 297) as follows: Put a minute drop of the albumen on the center of a slide and with a clean finger spread the albumen over the slide, wiping off all that is possible. Finally beat or tap the slide with the end of the finger. This will make a very thin (it cannot be too thin) and even layer. Place the sections in position and allow them to remain until the water has all evaporated. It is well to leave them over night. After the water has evaporated, coat the sections with ¾ ths collodion, using a delicate brush for the purpose. Allow the collodion to dry for a minute or two, then put the slides in benzin or xylene to dissolve the paraffin (see § 276). If the sections are not extended on water, they may be put directly on the albumenized slides, pressed down with the finger and coated with collodion. This is much more rapid, but does not get rid of the fine folds. See also the preface for albumenizing slides.

§ 276. **Removing the Paraffin.**—Immerse the slide in a vessel of xylene or benzin. This will dissolve the paraffin. An hour will usually suffice. One can hasten the solution of the paraffin by moving the slide in the solvent. In this way it may be dissolved in 5 to 10 minutes, or even less. It will do no harm to leave the slide in the benzin or xylene over night. Two or three days even might not do any harm, but it is usually better to proceed at once to the other operations.

§ 277. **Removing the Xylene or Benzin**—From the xylene or benzin plunge the slide bearing the sections into a jar of 95% alcohol, and leave it for a few minutes, or move it around in the alcohol for half a minute or so.

§ 278. **Staining the Sections with an Alcoholic Dye.**—With an alcoholic stain like hydrochloric acid carmine, remove the slide from the alcohol, and add the stain directly after draining the slide. Do not allow the stain to become dry, for that would injure the tissue. Wash away the stain with 67% alcohol, then dehydrate with 97% alcohol, clear and mount in balsam as described below.

§ 279. **Staining with an Aqueous Dye.**—Wash away the 95% alcohol from the slide bearing the sections by plunging it into a jar of water and moving it around a moment. Then add the stain to the sections with a pipette, or immerse the slide in a jar of the stain, and allow the stain to act from 5 to 10 minutes. Wash thoroughly with water.

§ 280. **Staining with a General Dye—Counterstaining.**—If it is desired to give a general stain after the nuclear dye (§ 279), carmine stained preparations may be tinted with picric-alcohol for half a minute or more (§ 315), and the hematoxylin stained specimens with eosin (§ 305). It usually takes less than a minute for this. Wash away the counterstain with water.

§ 280a. **Counterstaining with Picro-fuchsin.**—For a general dye to use with hematoxylin, eosin is good, but to differentiate the tissues more completely, especially connective tissue, which is present in practically every section made, it is better to use Van Gieson's mixture of picric acid and acid fuchsin. (Picric acid, saturated aqueous solution 75 cc., water 25 cc. 1% aqueous solution of acid fuchsin, 10 cc.) Sections are first strongly stained with hematoxylin, well washed with water, and then stained 3 seconds to 15 minutes in the picro-fuchsin. They are then washed in distilled water or in tap water, to which has been added a drop or two of glacial acetic acid to 100 cc. of the water. They are then dehydrated, cleared and mounted in acid balsam, that is in balsam which has not been neutralized (§ 300). If glycerin or glycerin jelly is used as a mounting medium it should be slightly acid. Unless the mounting medium is slightly acid, the red of the acid fuchsin soon fades. In some cases less acid fuchsin should be used, and in some a greater amount. Acid fuchsin alone without the picric acid is also good for a counterstain. The picro-fuchsin is a very valuable differential stain and combined in different proportions with picric acid will give great assistance in almost every case. It does not seem to be a very permanent stain. (See Freeborn, Trans. N. Y. Path. Soc., 1893, p. 73. Also studies from the department of pathology of the College of Physicians and Surgeons, Columbia University, N. Y., 1894-1895).

§ 281. **Dehydration of the Stained Sections.**—Place the slide with the stained sections in a jar of 95% alcohol and leave it a few minutes, or wave it around in the alcohol for half a minute or so.

§ 282. **Clearing the Sections.**—Drain off the alcohol, and place the slide in a jar of clearer (§ 303, A or B) or put a drop or two of clearer on the sections. The clearing is usually accomplished in two or three minutes.

§ 283. **Mounting in Balsam.**—For this the clearer is drained from the slide, and wiped away with lens or blotting paper, cloth, etc. The balsam is then put upon the sections and the cover added, or a cover-glass is spread with the balsam and put over the sections. (If the sections show a whitish appearance and are opaque they were not sufficiently dehydrated.)

ORDER OF PROCEDURE IN MAKING MICROSCOPICAL PREPARATIONS BY THE PARAFFIN METHOD.

§ 284. It will be seen from this table and from sections 268 to 283 that it requires from 5 to 7 days to get a microscopical preparation by the paraffin method if one starts with a fresh tissue. Depending on the method of fixing and hardening, the time may be much greater. Unless mush time is lost in waiting one must plan ahead in histological work.

1. Fixing and hardening the tissue or organ (§ 269), 4 days or more.
2. Dehydrating the object to be cut in 95% or stronger alcohol (§ 270), 1 to 24 hours.
3. Displacing the alcohol and clearing tissues with cedar-wood oil. (See § 271), 2 to 24 hours.
4. Infiltrating the tissue with paraffin in the paraffin oven (§ 271), 2 to 24 hours.
5. Imbedding in paraffin (§ 272), 10 minutes.
6. Cutting the sections (§ 273), 10 minutes.
7. Extending the sections with warm water. (See § 274, 274a.)
8. Fastening the sections to a slide (§ 275), 5 minutes to 24 hours.
9. Removing the paraffin (§ 276), 10 minutes to 24 hours.
10. Removing the xylene or benzin (§ 277).
11. Washing with water, note, p. 162.
12. Staining with an aqueous dye (§ 279), 2 minutes to 24 hours.
13. Washing away the superfluous stain with water (§ 279).
14. Staining with a general dye (§ 280–280a), 10 seconds to 10 minutes.
15. Washing the sections with water (§ 280–280a).
16. Dehydrating the stained sections in 95% alcohol (§ 281), 3 minutes to 24 hours.
17. Clearing the sections (§ 282), 2 minutes to 24 hours.
18. Mounting in Balsam (§ 283), 1 to 5 minutes.
19. Sealing the cover-glass (§ 238), 2 minutes.
20. Labeling the preparation (§ 292), 2 minutes.
21. Cataloging the preparation (§ 294), 5 to 10 minutes.

SERIAL SECTIONS.

§ 285. In histological studies it is frequently of the greatest advantage to have the sections in serial order, then an obscure feature in one section is frequently made clear by the following or preceding sections. While serial sections are very desirable in histological study, they are absolutely necessary for the solution of morphological problems presented in complex organs like the brain, in embryos and in minute animals where gross dissection is impossible.

§ 286. **Arrangement of Tissues for Sections in Histology.**—They should be so arranged that the exact relations of each part to the organ can be readily determined. For example, an organ like the intestine, a muscle or a nerve, should be so arranged that exact transections or longisections can be made. Organs like the liver and other glands, the skin, etc., should be so arranged that sections parallel with the surface or at right angles to it, (surface or vertical sections) may be made. Oblique sections are often very puzzling.

With cylindrical objects, especially botanical specimens, one may cut tangential sections, *i. e.*, sections at right angles to a radius, or parallel with the radii (radial sections), or transections, *i. e.*, sections across the long axis.

§ 287. **Arrangement of Serial Sections.**—The numerical order may be very conveniently like the words on a printed page, from the upper left hand corner and extending from left to right, top to bottom (Fig. 135).

The position of the various aspects of the sections should be in general such that when they are under the compound microscope the rights and lefts will correspond with those of the observer. This may be accomplished as follows for sections made in the three cardinal sectional planes, *Transections, Frontal Sections, Sagittal Sections:*

(A) *Transections, i. e.*, sections across the long axis of the embryo or animal dividing it into equal or unequal cephalic and caudal parts.

(a) In accordance with the generally approved method of numbering serial parts in anatomy, the most cephalic section should be first (No. 1 of Fig. 135).

(b) The caudal aspect of the section should face upward toward the cover-glass, the cephalic aspect being next the slide.

(c) The ventral aspect should face toward the upper edge of the slide (Fig. 135).

This arrangement may be easily accomplished in transections in two ways: (1) The embryo or animal is imbedded in such a way that the sectioning shall begin at the cephalic end. In this case the first section is placed in the upper left hand corner of the slide (No. 1 of Fig. 135), but it must be turned over so that the caudal aspect shall face up. The ventral aspect must be made to look toward the upper edge of the slide, then under the compound microscope the dorsal side will appear toward the upper edge of the slide and the right and left correspond with the observer.

(2) The embryo or animal is imbedded so that the sectioning begins at the caudal end, then the sections are not turned over, as they are already caudal face up, but they must be put on the slide in reverse order, *i. e.*, the first section made is put in the lower right hand corner (No. 10 of Fig. 135). In this way the most cephalic section will be number one as before. As in the previous case the ventral side of the section should be toward the upper edge of the slide (Fig. 135).

B) *Frontal sections, i. e.*, sections lengthwise of the embryo or animal and from right to left (dextral and sinistral), so that it is divided into equal or unequal dorsal and ventral parts.

The embryo is so imbedded and arranged in the microtome that the dorsal part is cut first. The first section is then placed in the upper left hand corner (No. 1, Fig. 135) dorsal side up, and the cephalic end toward the lower edge of the slide. The microscopic image will then appear with right and left as in the observer.

C) *Sagittal sections, i. e.*, sections lengthwise of the embryo or animal, and from the ventral to the dorsal side, thus dividing it into equal or unequal right and left parts. For these the left side of the embryo is placed up so that it is cut first. The first section is placed in the upper left hand corner of the slide, the left aspect facing away from the slide and the head to the right end, the ventral side toward the upper edge of the slide. Under the microscope the dorsal side will then appear toward the upper edge of the slide and the head to the left.

§ 288. For serial sections with collodion imbedded objects it is a great advantage to have the imbedding mass unsymmetrically trimmed, so that if a section is accidentally turned over it may be easily noticed and rectified.

Furthermore it is imperatively necessary that the object be so imbedded that the cardinal aspects, dextral and sinistral, dorsal and ventral, cephalic and caudal, shall be known with certainty.

UPPER EDGE OF SLIDE.

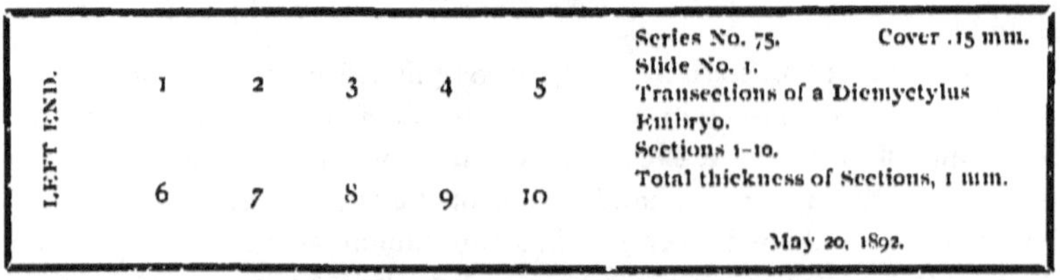

LEFT END.	1	2	3	4	5	Series No. 75. Cover .15 mm. Slide No. 1. Transections of a Diemyctylus Embryo. Sections 1–10. Total thickness of Sections, 1 mm.
	6	7	8	9	10	May 20, 1892.

FIG. 135.—*Labeled Slide of Serial Sections.*

§ 289. **Thickness of Cover-Glass and of Serial Sections.**—It is a great advantage to use very thin cover-glasses ($\frac{1.0}{100}$ to $\frac{1.5}{100}$ mm.) for serial sections, then the cover will not prevent the use of high powers. When the ordinary slides (25 × 76 mm., 1 × 3 inch) are used cover-glasses 23 × 55 mm. may be advantageously employed.

The combined thickness of the sections on a slide is easily determined by noting carefully the position of the microtome screw at the first and last sections, and measuring the elevation. Then if the sections are uniform the thickness of each may be easily found. The average thickness may be easily determined in any case.

§ 290. **Labeling Serial Sections.**—The label of a slide on which serial sections are mounted should contain at least the following : (1) The number of the series ; (2) The number of the slide in the series (if the series required more than one slide) ; (3) Kind of sections (transections, etc.) and the name of the object from which derived ; (4) The number of the first and last section on the slide ; (5) The total thickness of all the sections on the slide ; (6) The date of the series.

LABELING, CATALOGING AND STORING MICROSCOPICAL PREPARATIONS.

§ 291. Every person possessing a microscopical preparation is interested in its proper management ; but it is especially to the teacher and the investigator that the labeling, cataloging and storing of microscopical preparations are of importance. "To the investigator, his specimens are the most precious of his possessions, for they contain the facts which he tries to interpret, and they remain the same while his knowledge, and hence his power of interpretation, increase. They thus form the basis of further or more correct knowledge ; but in order to be safe guides for the student, teacher, or investigator, it seems to the writer that every preparation should possess two things : viz., a label and a catalog or history. This catalog should indicate all that is known of a specimen at the time of its preparation, and all of the processes by which it is treated. It is only by the possession of such a complete knowledge of the entire history of a preparation that one is able to judge with certainty of the comparative excellence of methods, and thus to discard or improve those which are defective. The teacher, as well as the investigator, should have this information in an accessible form, so that not only he, but his students can obtain at any time, all necessary information concerning the preparations which serve him as illustrations and them as examples."

§ 292. **Labeling Ordinary Microscopical Preparations.**—The label should possess at least the following information (see § 290 for serial sections) :

EXAMPLE.

(1) The number of the preparations, the thickness of the cover-glass and of the sections under it.
No. 475. C. .15
Secs. 8 u

(2) The name and source of the preparation.
Striated Muscle ; transection of the Sartorius of the Cat.

(3) The date of the specimen (2 of catalog.)
October 15, 1894.

§ 293. **Cataloging Preparations.**—It is believed from personal experience, and from the experience of others, that each preparation (each slide or each series) should be accompanied by a catalog containing at least the information suggested in the following formula: This formula is very flexible, so that the order may be changed, and numbers not applicable in a given case may be omitted. With many objects, especially embryos and small animals, the time of fixing and hardening may be months or even years earlier than the time of imbedding. So, too, an object may be sectioned a long time after it was imbedded, and finally the sections may not be mounted at the time they are cut. It would be well in such cases to give the date of fixing under 2, and under 5, 6 and 8, the dates at which the operations were performed if they differ from the original date and from one another. In brief, the more that is known about a preparation the greater its value.

§ 294. **General Formula for Cataloging Microscopical Preparations:**

1. The general name and source. Thickness of cover glass and of sections.

2. The number of the preparation and the date of obtaining and fixing the specimen; the name of the preparator.

3. The special name of the preparation and the common and scientific name of the object from which it is derived. Purpose of the preparation.

4. The age and condition of the object from which the preparation is derived. Condition of rest or activity; fasting or full fed at the time of death.

5. The chemical treatment, — the method of fixing, hardening, dissociating etc., and the time required.

6. The mechanical treatment, — imbedded, sectioned, dissected with needles, etc. Date at which done.

7. The staining agent or agents and the time required for staining.

8. Dehydrating and clearing agent, mounting medium, cement used for sealing.

9. The objectives and other accessories (micro-spectroscope, polarizer, etc.) for studying the preparation.

10. Remarks, including references to original papers, or to good figures and descriptions in books.

A Catalog Card Written According to this Formula:

Muscular Fibers. Cat. C. 15.
Fibers 20 to 40 μ thick..

2. No. 475. (Drr. IX) Oct. 1, 1891. S. H. G., Preparator.

3. Tendinous and intra-muscular terminations of striated muscular fibers from the *Sartorius* of the cat (*Felis domestica*.)

4. Cat eight months old, healthy and well nourished. Fasting and quiet for 12 hours.

5. Muscle pinned on cork with vaselined pins and placed in 20 per cent. nitric acid immediately after death by chloroform. Left 36 hours in the acid; temperature 20° C. In alum water (½ sat. aq. sol.) 1 day.

6. Fibers separated on the slide with needles. Oct. 3.

7. Stained 5 minutes with Delafield's hematoxylin.

8. Dehydrated with 95%. alcohol 5 minutes, cleared 5 minutes with carbol-turpentine, mounted in xylene balsam; sealed with shellac.

9. Use 18 mm. for the general appearance of the fibers, then 2 or 3 mm. objective for the details of structure. Try the micro-polariscope (§ 209).

10. The nuclei or muscle corpuscles are very large and numerous; many of the intra-muscular ends are branched. See S. P. Gage, Proc. Amer. Micr. Sci., 1890, p. 132; Ref. Hand-book Med., Sci., Vol. V., p. 59.

§ 295. **General Remarks on Catalogs and Labels.**—It is especially desirable that labels and catalogs shall be written with some imperishable ink. Some form of water-proof carbon ink is the most available and satisfactory. The water-proof India ink, or the engrossing carbon ink of Higgins, answers very well. As purchased, the last is too thick for ordinary writing and should be diluted with one-third its volume of water and a few drops of strong ammonia added.

If one has a writing diamond it is a good plan to write a label with it on one end of the slide. It is best to have the paper label also, as it can be more easily read.

FIG. 136. *Writing diamond for writing numbers and labels on glass slides, cutting cover-glasses, etc. (Queen & Co.).*

The author has found stiff cards, 12½x7½ cm., like those used for cataloging books in public libraries, the most desirable form of catalog. A specimen that is for any cause discarded has its catalog card destroyed. New cards may then be added in alphabetical order as the preparations are made. In fact a catalog on cards has all the flexibility and advantages of the slip system of notes (see Wilder & Gage, p. 45).

Some workers prefer a book catalog. Very excellent book catalogs have been devised by Alling and by Ward (Jour. Roy. Micr. Soc., 1887, pp. 173, 348 ; Amer. Monthly Micr. Jour., 1890, p. 91 ; Amer. Micr. Soc. Proc., 1887, p. 233).

The fourth division has been added as there is coming to be a very strong belief, practically amounting to a certainty, that there is a different structural appearance in many if not all of the tissue elements depending upon the age of the animal, upon its condition of rest or fatigue ; and for the cells of the digestive organs, whether the animal is fasting or full fed. Indeed as *physiological histology* is recognized as the only true histology, there will be an effort to determine exact data concerning the animal from which the tissues are derived. (See Minot, Proc. Amer. Assoc. Adv. Science, 189 , pp. 271-289 ; Hodge, on nerve cells in rest and fatigue, Jour. Morph., vol. VII. (1892), pp. 95-168 ; Jour. Physiol., vol. XVII., pp. 129-134 ; Gage, The processes of life revealed by the microscope ; a plea for physiological histology, Proc. Amer. Micr. Soc., vol. XVII. (1895), pp. 3-29 ; Science. vol. II., Aug. 23, 1895, pp. 209-218).

CABINET FOR MICROSCOPICAL PREPARATIONS.

§ 296. While it is desirable that microscopical preparations should be properly labeled and cataloged, it is equally important that they should be protected from injury. During the last few years several forms of cabinets or slide holders have been devised. Some are very cheap and convenient where one has but a few slides. For a laboratory or for a private collection where the slides are numerous the following characters seem to the writer essential :

(1). The cabinet should allow the slides to lie flat, and exclude dust and light.

(2). Each slide or pair of slides should be in a separate compartment. At each end of the compartment should be a groove or bevel, so that upon depressing either end of the slide the other may be easily grasped (Fig. 140). It is also desirable to have the floor of the compartment grooved so that the slide rests only on two edges, thus preventing soiling the slide opposite the object.

(3). Each compartment or each space sufficient to contain one slide of the standard size should be numbered, preferably at each end. If the compartments are made of sufficient width to receive two slides, then the double slides so frequently used in mounting serial sections may be put into the cabinet in any place desired.

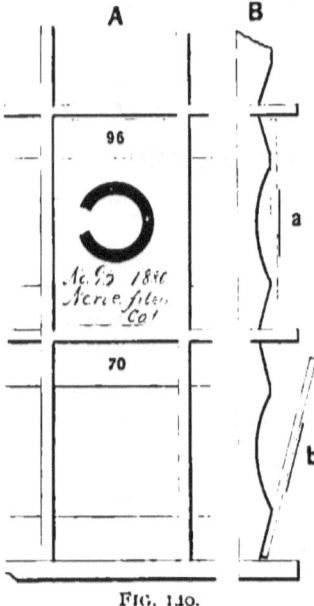

Fig. 140.

Fig. 140.—*A*—. *Part of a cabinet drawer seen from above. In compartment No. 96 is represented a slide lying flat. The label of the slide and the number of the compartment are so placed that the number of the compartment may be seen through the slide. The sealing cement is removed at one place to show that in sealing the cover-glass, the cement is put partly on the cover and partly on the slide. (§ 229, 234).*

B.—This represents a section of the same part of the drawer. (a) Slide resting as in A. No. 96 The preparation is seen to be above a groove in the floor of the compartment. (b) One end of the slide is seen to be uplifted by depressing the other into the bevel.

(4). The drawers of the cabinet should be entirely independent, so that any drawer may be partly or wholly removed without disturbing any of the others.

(5). On the front of each drawer should be the number of the drawer in Roman numerals, and the number of the first and last compartment in the drawer in Arabic numerals. (Fig. 141).

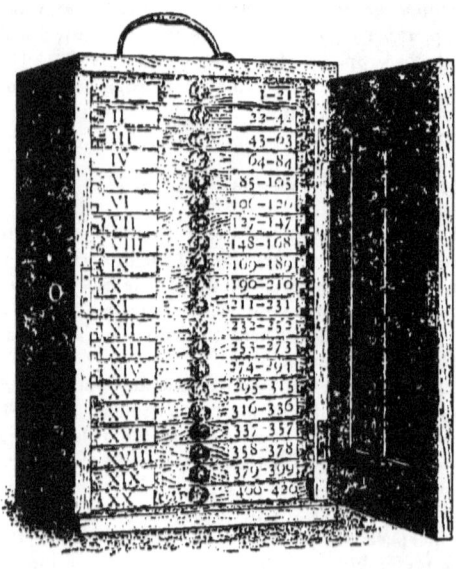

Fig. 141.—*Cabinet for Microscopical Specimens, showing the method of arrangement and of numbering the drawers and indicating the number of the first and last compartment in each drawer. It is better to have the slides on which the drawers rest somewhat shorter, then the drawer front may be entire and not notched as here shown.* (From Proc. Amer. Micr. Soc. 1883).

REAGENTS FOR FIXING, MOUNTING, ETC.

§ 297. **Albumen Fixative (Mayer's).**—This consists of equal parts of well-beaten white of egg and glycerin. To each 50 cc. of this 1 gram of salicylate of soda is added to prevent putrefactive changes. Probably a small amount of formaldehyde, say 1 cc. of the 40%, to 50 or 100 cc. of the fixative would suffice; if too strong the albumen would be precipitated. For method of use see § 275.

§ 298. **Alcohol (Ethylic)**—Ethyl alcohol is mostly used for histological purposes. (A) Absolute alcohol (i. e. alcohol of 99-100%) is recommended for many purposes, but if plenty of 95% alcohol is used it answers every purpose in histology.

(B) 82% alcohol made by mixing 5 parts of 95% alcohol with 1 part of water.

(C) 67% alcohol made by mixing 2 parts of 95% alcohol with 1 part of water.

§ 299. **Alum Solution.**—For muscle dissociated in nitric acid (§ 249) a saturated solution (i. e. a solution in which the water holds all the alum it can. If one adds an excess so that there will always be some undissolved alum in the vessel he can be sure the solution is saturated after it has stood a few days. An easy way to get a saturated solution is to take 500 cc. of water and add 100 grams of alum and heat the water in an agate dish. All the alum will be melted, but on cooling a part will crystallize out, leaving a cold saturated solution). The saturated solution may be used but, if a half saturated solution is employed, it will answer all the purposes. For a half saturated solution take 100 cc. of water and 100 cc. of saturated alum water and mix the two.

§ 300. **Balsam, Canada Balsam, Balsam of Fir; Xylene Balsam.**—This is one of the oldest and most satisfactory of the resinous media used for mounting microscopical preparations. Sometimes it is used in the natural state, but experience has shown that it is better to get rid of the natural volatile constituents. A considerable quantity, half a liter or more, of the natural balsam is poured into shallow plates in layers about 1 or 2 centimeters thick, then the plates are put in a warm, dry place, on the back of a stove or on a steam radiator, and allowed to remain until the balsam may be powdered when it is cold. This requires a long time, the time depending on the temperature and the thickness of the layer of balsam. By heating the natural balsam in a tin or agate vessel over a Bunsen burner or an alcohol lamp the time may be greatly abbreviated. The heat should not be sufficient to boil the balsam, however.

When the volatile products have evaporated, the balsam is broken into small pieces or powdered in a mortar and mixed with about an equal volume of xylene, turpentine or chloroform. It will dissolve in these, and then should be filtered through absorbent cotton or a filter paper, using a paper funnel.* The balsam is too thin in this condition for mounting, but so made for the sake of filtering it. After it is filtered it is evaporated slowly in an open dish or a wide-mouth bottle or jar till it is of a syrupy consistency at the ordinary temperature. It is then poured into a bottle with a glass cap like a spirit lamp. For use it is put into a small spirit lamp (Fig. 137).

---

*For filtering balsam and all resinous and gummy materials, the writer has found a paper funnel the most satisfactory. It can be used once and then thrown away. Such a funnel may be very easily made by rolling a sheet of thick writing paper in the form of a cone and cementing the paper where it overlaps, or winding a string several times around the lower part. Such a funnel is best used in one of the rings for holding funnels.

The xylene is much the best substance to use for thinning the balsam. Such xylene balsam, as it is then called, may be used for mounting any object suitable for balsam mounting. The dehydration must be very perfect, however, as xylene is wholly immiscible with water.

Natural balsam is liable to be slightly acid. This is of advantage for mounting sections stained with carmine or injected with carmin gelatin or Berlin blue gelatin. For hematoxylin preparations and for fuchsin preparations the acid will cause the color to fade. The balsam may be neutralized by mixing some carbonate of soda with the thinned solution before it is thickened. In a few days all the soda will settle and the clear balsam above will be neutral and may be poured off and thickened. If one mounts carmine or Berlin blue preparations in the neutral balsam the blue will fade and the carmine diffuse.

§ 301. **Chloroform Paraffin.**—This is made by mixing the 4 parts of the paraffin used for imbedding (§ 314) with 1 part of chloroform. This gives a paraffin which melts at a lower temperature than the pure paraffin. If it is kept warm the chloroform evaporates in 3 to 6 days, leaving pure paraffin.

§ 302. **Clarifier, Castor-Xylene Clarifier.**—This is composed of castor oil 1 part and xylene* 3 parts.

§ 303. **Clearing Mixture** (§ 266, 282).—(A). One of the most satisfactory and generally applicable clearers is carbol turpentine, made by mixing carbolic acid crystals (*Acidum carbolicum. A. phenicum crystalizatum*) 40 cc. with rectified oil of turpentine (*Oleum terebinthinae rectificatum*) 60 cc. If the carbolic acid does not dissolve in the turpentine add 5 cc. of 95% alcohol, or increase the turpentine, thus : carbolic acid 30 cc., turpentine 70 cc.

(B). **Carbol-Xylene, Clearer.**—Vasale recommends as a clearer, xylene 75 cc., carbolic acid (melted crystals) 25 cc. It is used in the same way as the preceding.

§ 304. **Collodion.**—This is a solution of soluble cotton† or other form of pyroxylin in equal parts of sulphuric ether and 95% alcohol. Three solutions are used :

* The hydrocarbon xylene (C_8H_{10}) is called xylol in German. In English, members of the hydrocarbon series have the termination "ene" while members of the alcohol series terminate in "ol."

† The substance used in preparing collodion goes by various names, soluble cotton or collodion cotton is perhaps best. This is cellulose nitrate, and consists of a mixture of cellulose tetranitrate $C_{12}H_{16}(NO_3)_4O_6$, and cellulose pentanitrate, $C_{12}H_{15}(NO_3)_5O_5$. Besides the names soluble and collodion cotton, it is called gun cotton and pyroxylin. Pyroxylin is the more general term and includes several of the cellulose nitrates. Celloidin is a patent preparation of pyroxylin, more expensive than soluble cotton, but in no way superior to it for imbedding.

Soluble cotton should be kept in the dark to avoid decomposition. After it is in solution this decomposition is not so liable to occur. The decomposition of the dry cotton gives rise to nitrous acid, and hence it is best to keep it in a box loosely covered so that the nitrous acid may escape.

Cellulose nitrate is explosive under concussion and when heated to 150° centigrade. In the air, the loose soluble cotton burns without explosion. It is said not to injure the hand if held upon it during ignition and that it does not fire gunpowder if burned upon it. So far as known to the writer, no accident has ever occurred from the use of soluble cotton for microscopical purposes. I wish to express my thanks to Professor W. R. Orndorff, organic chemist in Cornell University, for the above information. Proc. Amer. Micr. Soc., vol. XVII (1895), pp. 361-370.

(A). 6% or *thick collodion*. It is made by mixing 50 cc. of sulphuric ether and 50 cc. of 95% alcohol and adding 6 grams of soluble cotton. If this is shaken repeatedly the solution will be complete in a day or two.

(B). 1½% or *thin collodion*. To prepare this 1½ grams of soluble cotton are added to 100 cc. of ether-alcohol (§ 306).

(C). ¼% collodion or *cementing collodion*. To prepare it ¼ ths of a gram of soluble cotton is added to 100 cc. of ether-alcohol.

As both ether and alcohol are very volatile it is necessary to keep the bottles containing them well corked.

§ 305. Eosin.—This is used mostly as a contrast stain with hematoxylin, which is an almost purely nuclear stain. It serves to stain the cell-body, ground substance, etc., which would be too transparent and invisible with hematoxylin alone. If eosin is used alone it gives a decided color to the tissue and thus aids in its study (§ 135). Eosin is used in alcoholic and in aqueous solutions. A very satisfactory stain is made as follows : 50 cc. of water and 50 cc. of 95% alcohol are mixed and 1-10th of a gram of dry eosin added.

The eosin is used after the hematoxylin in most cases (§ 280), and, as it is in alcoholic solution, it may be washed off with 95% alcohol if the object is to be mounted in balsam. If it is to be mounted in glycerin or glycerin jelly, the excess of eosin should be washed away with distilled water.

§ 306. Ether, Ether-Alcohol.—Sulphuric ether is meant when ether is mentioned in this book. For the ether-alcohol mentioned in § 254, 304, etc., a mixture of equal volumes of sulphuric ether and 95% alcohol is meant.

§ 307. Farrant's Solution.—Take 25 grams of clean, dry, gum arabic ; 25 cc. of a saturated aqueous solution of arsenious acid ; 25 cc. of glycerin. The gum arabic is soaked for several days in the arsenic water, then the glycerin is added and carefully mixed with the dissolved or softened gum arabic.

This medium retains air bubbles with great tenacity. It is much easier to avoid than to get rid of them in mounting.

§ 308. Formaldehyde Dissociator.—This is composed of 5 cc. of a 40% solution of formaldehyde in 995 cc. of water, to which 6 grams of common table salt (sodium chlorid) have been added. That is, it is a $\frac{1}{5}$% solution of formaldehyde in normal salt solution (§ 313). Formaldehyde as bought in the market is a 40% solution in water, and is called formol, formalin, formalose and formal, the last name being the preferable one. For its use in isolating cells see § 245. (Gage Micr. Bulletin and Sci. News, vol. XII. (1895), pp. 4-5).

§ 309. Glycerin.—(A). One should procure pure glycerin for a mounting medium. It needs no preparation, except in some cases it should be filtered through filter paper or absorbent cotton to remove dust, etc.

For preparing objects for final mounting, glycerin 50 cc., water 50 cc., forms a good mixture. For many purposes the final mounting in glycerin is made in an acid medium, viz., Glycerin 99 cc., Glacial acetic or formic acid, 1 cc.

By extreme care in mounting and by occasionally adding a fresh coat to the sealing of the cover-glass, glycerin preparations last a long time. They are liable to be very disappointing, however. In mounting in glycerin care should be taken to avoid air-bubbles, as they are difficult to get rid of. A specimen need not be discarded, however, unless the air-bubbles are large and numerous.

Glycerin Jelly.—Soak 25 grams of the best dry gelatin in cold water in a small agate-ware dish. Allow the water to remain until the gelatin is softened. It usually takes about half an hour. When the gelatin is softened, as may be readily determined by taking a little in the fingers, pour off the superfluous water and drain well to get rid of all the water that has not been imbibed by the gelatin. Warm the softened gelatin over a water bath and it will melt in the water it has absorbed. Add to the melted gelatin about 5 cc. of egg albumen, white of egg ; stir it in well and then heat the gelatin in the water bath for about half an hour. Do not heat above 75° or 80° C., for if the gelatin is heated too hot it will be transformed into meta-gelatin and will not set when cold. The heat will coagulate the albumen and form a kind of floculent precipitate which seems to gather all fine particles of dust, etc., leaving the gelatin perfectly clear. After the gelatin is clarified it should be filtered through a hot flannel filter and mixed with an equal volume of glycerin and 5 grams of chloral hydrate and shaken thoroughly. If it is allowed to remain in a warm place (*i. e.*, in a place where the gelatin remains melted) the air-bubbles will rise and disappear.

In case the glycerin jelly remains fluid or semi-fluid at the ordinary temperature (18°-20° C.), the gelatin has either been transformed into meta-gelatin by too high a temperature or it contains too much water. The amount of water may be lessened by heating at a moderate temperature over a waterbath in an open vessel. This is a very excellent mounting medium. Air-bubbles should be avoided in mounting as they do not disappear.

§ 310. **Hematoxylin.**—Hematoxylin is one of the most useful stains employed in histology. A very excellent solution for ordinary section staining may be made as follows : Distilled water 200 cc., and potash alum $7\frac{1}{2}$ grams, are boiled together for 5 minutes, in an agate-ware or glass vessel, and sufficient boiled water added to bring the volume back to 200 cc. After the mixture is cool, 4 grams of chloral hydrate, and $\frac{2}{10}$ths gram of hematoxylin crystals, previously dissolved in 20 cc. of 95% alcohol, are added. The boiling seems to destroy any fungi present in the alum or water, and the chloral prevents the development of any that may get in afterward, and this solution therefore is quite permanent.

At first the color will be rather faint, but after a week or two it will become a deep purple. The deepening of the color is more rapid if the bottle is left uncorked in the light and is shaken occasionally.

If the stain is too concentrated it may be diluted with freshly distilled water or with a mixture of water, alum and chloral. If the stain is not sufficiently concentrated, more hematoxylin may be added. With hematoxylin of the strength given in the formula, sections are usually sufficiently stained in from one to five minutes.

As may be inferred from what was said above, the boiling is to destroy any living ferments present in the water or alum, and the chloral hydrate is to prevent the development of germs which accidentally reach the solution after it is made.

No precaution is necessary in using this stain for sections, except that applicable to all hematoxylin solutions, viz. : after staining, the surplus stain must be very thoroughly washed away with distilled water ; otherwise black granules or needles will appear in or upon the sections. If granules appear in the preparations in spite of the washing, it will be well to boil the solution three to five minutes and filter through paper or absorbent cotton. The addition of one or two per cent. of chloral after the boiling is also advantageous. This stain has not been tried for dyeing in bulk. Other substances than chloral were tried, but not with so good success. (S. H. Gage, Proc. Amer. Micr. Soc., Vol. XIV, 1892, pp. 125-127).

§ 311. **Liquid Gelatin.**—Gelatin or clear glue, 75 to 100 grams. Commercial acetic acid (No. 8) 100 cc., water 100 cc., 95% alcohol 100 cc., glycerin 15 to 30 cc. Crush the glue and put it into a bottle with the acid, and set in a warm place, and shake occasionally. After three or more days add the other ingredients. This solution is excellent for fastening paper to glass, wood or paper. The brush must be mounted in a quill or wooden handle. For labels, it is best to use linen paper of moderate thickness. This should be coated with the liquid gelatin and allowed to dry. The labels may be cut of any desired size and attached by simply moistening them, as in using postage stamps.

Very excellent blank labels are now furnished by dealers in microscopical supplies, so that it is unnecessary to prepare them one's self, except for special purposes.

§ 312. **Nitric Acid Dissociator.**—This is prepared by mixing 80 cc. of water with 20 cc. of strong nitric acid. It is used mostly in dissolving the connective tissue of muscle and thus making it possible to separate the fibers. Alum water is used as a restrainer (§ 299 and 249). (Gage, Proc. Amer. Micr. Soc., Vol. XI, (1889), pp. 34–45).

§ 313. **Normal Salt Solution or Saline Solution.**—Pure water from its differing density from the natural lymph acts injuriously on the tissues. The addition of a little table salt, however, prevents this deleterious action, or greatly lessens it, hence the name of *normal salt solution*. It is a $\frac{6}{10}$% solution of table salt (sodium chlorid) in water; water 1000 cc., salt 6 grams, or water 100 cc., salt $\frac{6}{10}$ gram.

§ 314. **Paraffin.**—Paraffin is of various melting points, hence at the ordinary temperature of a laboratory, that melting at the lowest temperature will be moderately soft, hence *soft paraffin*, while that melting at a higher temperature will be hard. For the best results one has to mix hard and soft paraffins. Usually a mixture of 9 parts hard and 1 part soft paraffin will give good results, and may be called *imbedding paraffin*. For *chloroform paraffin*, 4 parts of imbedding paraffin are mixed with one part of chloroform (§ 301, 272).

§ 315. **Picric-Alcohol.**—This is an excellent hardener and fixer for almost all tissues and organs. It is composed of 500 cc. of water and 500 cc. of 95% alcohol, to which 2 grams of picric acid have been added. (It is a $\frac{1}{5}$% solution of picric acid in 50% alcohol). It acts quickly, in from one to three days. (§ 252, 269). (Proc. Amer. Micr. Soc., Vol. XII (1890), pp. 120–122).

§ 316. **Shellac Cement.**—Shellac cement for sealing preparations and for making shallow cells (§ 233) is prepared by adding scale or bleached shellac to 95% alcohol. The bottle should be filled about half full of the solid shellac then enough 95% alcohol added to fill the bottle nearly full. The bottle is shaken occasionally and then allowed to stand until a clear stratum of liquid appears on the top. This clear, supernatant liquid is then filtered through absorbent cotton, using a paper funnel (§ 300, note), into an open dish or a wide-mouth bottle. To every 50 cc. of this filtered shellac, 5 cc. of castor oil and 5 cc. of Venetian turpentine are added to render the shellac less brittle. The filtered shellac will be too thin, and must be allowed to evaporate till it is of the consistency of thin syrup. It is then put into a capped bottle, and for use, into a small spirit lamp (Fig. 134). In case the cement gets too thick add a small amount of 95% alcohol or some thin shellac. The solution of shellac almost always remains muddy, and in most cases it takes a very long time for the flocculent substance to settle. One can very quickly obtain a clear solution as follows: When the shellac has had time to thoroughly dissolve, *i. e.*, in a week or two in a warm place, or in less time if the bottle is frequently shaken, a part of the dissolved shellac is poured into a bottle and about one fourth as much gasolin or benzin added and the two well shaken. After twenty-four hours or so the flocculent, undissolved substance will separate from the shellac solution and rise with the benzin to the top. The clear solution may then be siphoned off or drawn off from the bottom if one has an aspirating bottle. (R. Hitchcock, Amer. Monthly Micr. Jour., July, 1884, p. 131).

ARRANGING AND MOUNTING MINUTE OBJECTS.

§ 317. Minute objects like diatoms and the scales of insects may be arranged in geometrical figures or in some fanciful way, either for ornament or more satisfactory study. To do this the cover-glass is placed over the guide. This guide for geometrical figures may be a net-micrometer or a series of concentric circles. In order that the objects may remain in place, however, they must be fastened to the cover-glass. As an adhesive substance, liquid gelatin (§ 311) thinned with an equal volume of 50% acetic acid answers well. A very thin coating of this is spread on the cover with a needle, or in some other way, and allowed to dry. The objects are then placed on the gelatinized side of the cover and carefully got into position with a mechanical finger, made by fastening a cat's whisker in a needle holder. For most of these objects a simple microscope with stand (Figs. 19, 20, 130, 131) will be found of great advantage. After the objects are arranged, one breathes very gently on the cover-glass to soften the gelatin. It is then allowed to dry and if a suitable amount of gelatin has been used, and it has been properly moistened, the objects will be found firmly anchored. In mounting, one may use Canada balsam or mount dry on a cell (§ 232, 240). See Newcomer, Amer. Micr. Soc.'s Proc., 1885, p. 128; see also E. H. Griffith and H. L. Smith, Amer. Jour. of Micros., iv, 102, v, 87; Amer. Monthly Micr. Jour., i, 66, 107, 113. Cunningham, The Microscope, viii, 1888, p. 237.

MICRO-CHEMISTRY AND CRYSTALLOGRAPHY—EXPERIMENTS.

§ 318. The student of science, and especially chemistry, so frequently requires a knowledge of the appearance of minute crystals to aid in the determination of an unknown substance, or for his information in studying objects where crystals are liable to occur, that a few experiments have been introduced to give him a start in preparing and permanently mounting some of the common crystals.

It is recommended that the crystals be made in several ways, that is, from alcoholic solutions, aqueous solutions saturated and dilute, by spontaneous drying and crystallization and by rapid crystallization by the aid of heat. The modifications in crystallization under these different methods of treatment are frequently very striking.

In every case the student is advised to study the appearance of the crystals in the "mother liquor." As a rule, their characteristics are more clearly shown in the "mother liquor" than under any other conditions.

It is of very great advantage to examine all crystalline forms with polarized light (§ 209).

§ 319. **Determination of the Character of the Solid Sediment in Water.**—Take some of the sediment from a filter or allow a considerable volume of water to stand in a tall glass vessel to deposit its sediment. Take a concentrated drop of this sediment and mount it on a slide under a cover-glass. Study the preparation with the microscope. Probably there will be an abundance of animal and vegetable life as well as of solid sediment. Put a drop of dilute sulphuric acid (*Acidum sulphuricum, i. e.*, strong sulphuric acid 1 gram, water 9 grams) at the edge of the cover

FIG. 139. *Czapski's Ocular Iris-diaphragm with cross hairs for examining and accurately determining the axial images of small crystals. The iris diaphragm enables the observer to make the field as large or small as desired.*

A. Longitudinal section.

B. Transection, showing the cross lines and the iris diaphragm with the projecting part at the left, by which the diaphragm is opened and closed. (Zeiss' Catalog, No. 30).

and at the opposite edge a small piece of the lens paper (Fig. 128). The acid will gradually diffuse, and if the solid particles are carbonate of lime, minute bubbles will be seen to be given off. If they are silica or clay no change will result. Sulphuric acid is recommended for this, as the microscope would be far less liable to injury than as if some acid giving off fumes were used.

§ 320. **Herapath's Method of Determining Minute Quantities of Quinine.**—For a so-called test fluid 12 cc. of glacial acetic acid, 4 cc. of 95% alcohol, and 7 drops of dilute sulphuric acid (§ 319) are mixed. A drop of the test fluid is put on a slide and a very minute amount of quinine added. After this is dissolved, add an extremely minute drop of an alcoholic solution of iodine. "The first effect is the production of the yellow cinnamon-colored compound of iodine and quinine, which forms as a small circular spot; the alcohol separates in little drops, which, by a sort of repulsive movement, drive the fluid away; after a time the acid liquid again flows over the spot, and the polarizing crystals of sulphate of iodo-quinine are slowly produced in beautiful rosettes. This succeeds best without the application of heat." Dr. Herapath used this method to determine the presence of quinine in the urine of patients under quinine treatment. See Hogg, p. 150; Quarterly Jour. Micr. Sc., vol. ii, pp. 13–18. For further papers on micro-chemistry by Dr. Herapath, see the Royal Society's Catalog of Scientific Papers.

§ 321. **List of Substances for the Study of Crystallography with the Microscope.*** The substances are crystalized on the cover-glass in all cases, and in all cases, except where otherwise stated, a saturated aqueous solution of the substances was first prepared.

1. Ammonium chlorid; 2. Ammonium copper chlorid; 3. Barium chlorid; 4. Cobalt chlorid (beautiful crystals obtained by mixing the saturated aqueous solution with an equal volume of 95% alcohol). Crystallization in a current of dry

---

* Most of the chemicals here named were suggested to the writer by Prof. L. M. Dennis, of the Chemical Department.

air some distance above an alcohol or Bunsen flame; mount in xylene balsam (§ 300), or one may fuse the salt with the balsam (§ 300); 5. Copper acetate; Mount dry (§ 231); 6. Copper sulphate. Crystals much more satisfactory when examined in the "mother liquor." 7. Lead nitrate; 8. Mercuric chlorid (corrosive sublimate), mount in xylene balsam (§ 300, 240). 9. Nickel nitrate, obtain crystals by heating; mount in xylene balsam (§ 240, 300); 10. Potash alum; 11. Potassium chlorate; 12. Potassium dichromate. Compare specimen crystallized by heat and spontaneously; mount dry or in xylene balsam (§ 300). 13. Potassium iodide. Dilute with one or two volumes of water, and crystallize by heat. 14. Potassium nitrate; 15. Potassium oxalate; 16. Potassium sulphate; 17. Salicin. Fuse the dry salicin on the cover-glass, mount dry, or preferably fuse in balsam (§ 300). 18. Salicylic acid. Make a 10% solution in 95% alcohol; let it crystallize spontaneously in the air; mount dry (§ 231). 19. Sodium chlorid (common salt). Mix sat. aq. sol. with one or two volumes of water, and heat; mount dry or in balsam. 20. Sulphonal (§ 218).

§ 322. For directions and hints in micro-chemical work and crystallography, consult the various volumes of the Journal of the Roy. Micr. Soc.; Zeitschrift für physiologische Chemie, and other chemical journals; Wormly; Klément & Renard; Carpenter-Dallinger; Hogg; Behrens, Kossel und Schiefferdecker; Frey; Dana, and other works on mineralogy; Davis.

CHAPTER VIII.

PHOTO-MICROGRAPHY AND PHOTOGRAPHY WITH A VERTICAL CAMERA.*

APPARATUS AND MATERIAL FOR THIS CHAPTER.

Compound microscope with achromatic condenser; Achromatic and apochromatic object ives; Oculars, ordinary and projection; Lamp and bull's-eye condenser of some form; Photo micrographic camera, and an ordinary copying camera; Focusing glass; dry plates, developer, fixer, trays, dark room, and the other things needed for photography, like printing frames, etc., etc.; Photographic-objectives, one for large objects and one for small objects to be magnified from two to fifteen diameters.

§ 323. Nothing would seem more natural than that the camera, armed with a photographic objective or with a microscopic objective, should be called into the service of science to delineate with all their complexity of detail, the myriads of forms studied. Indeed, the very first pictures made on white paper and white leather, sensitized by silver nitrate, were made by the aid of a solar microscope (1802). The pictures were made by Wedgwood and Davy, and Davy says: "I have found that images of small objects produced by means of the solar microscope may be copied without difficulty on prepared paper." †

* Considerable confusion exists as to the proper nomenclature of photography with the microscope. In Germany and France the term micro-photography is very common, while in English photo-micrography and micro-photography mean different things. Thus: A *photo-micrograph* is a photograph of a small or microscopic object, usually made with a microscope and of sufficient size for observation with the unaided eye; while a *micro-photograph* is a small or microscopic photograph of an object, usually a large object, like a man or woman, and is designed to be looked at with a microscope.
Dr. A. C. Mercer, in an article in the Proc. Amer. Micr. Soc., 1886, p. 131, says that Mr. George Shadbolt made this distinction. See the Liverpool and Manchester *Photographic Journal* (now *British Journal of Photography*), Aug. 15, 1858, p. 203; also Sutton's Photographic Notes, Vol. III, 1858, pp. 205-208. On p. 208 of the last, Shadbolt's word "Photomicrography" appears. Dr. Mercer puts the case very neatly as follows: "A *photo micrograph* is a *macroscopic* photograph of a *microscopic* object; a *micro-photograph* is a *micro scopic* photograph of a *macroscopic* object." See also *Medical News*, Jan. 27, 1891, p. 108.
† In a most interesting paper by A. C. Mercer on "The Indebtedness of Photography to Microscopy, Photographic Times Almanac, 1887, it is shown that : "To

Thus among the very first of the experiments in photography the microscope was called into requisition. And naturally, plants and motionless objects were photographed in the beginnings of photography when the time of exposure required was very great.

At the present time photography is used to an almost inconceivable degree in all the arts and sciences and in pure art. Even astronomy finds it of the greatest assistance.

Although first in the field, Photo-Micrography has been least successful of the branches of photography. This is due to several causes. In the first place, microscopic objectives have been naturally constructed to give the clearest image to the eye, that is the visual image as it is sometimes called, is for microscopic observation, of prime importance. The actinic or photographic image, on the other hand, is of prime importance for photography. Then for the majority of microscopic objects transmitted light ($\frac{2}{3}$ 50) must be used, not reflected light as in ordinary vision. Finally, from the shortness of focus and the smallness of the lenses, the proper illumination of the object is accomplished with some difficulty, and the fact of the lack of sharpness over the whole field with any but the lower powers, have all combined to make photo-micrography less successful than ordinary macro-photography. So tireless, however, have been the efforts of those who believed in the ultimate success of photo-micrography, that now the ordinary achromatic objectives and ortho chromatic or isochromatic plates give very good results, while the apochromatic objectives with projection oculars give excellent results, even in hands not especially skilled. The problem of illumination has also been solved by the construction of achromatic and apochromatic condensers and by the electric and other powerful lights now available. There still remains the the difficulty of transmitted light and of so preparing the object that structural details shall stand out with sufficient clearness to make a picture which shall approach in definiteness the drawing of a skilled artist.

The writer would advise all who wish to undertake photo-micrography seriously, to study samples of the best work that has been produced. Among those who showed the possibilities of photo-micrographs was Col. Woodward of the U. S. Army Medical Museum. The photo-micrographs made by him and exhibited at the Centennial Celebration at Philadelphia in 1876, serve still as models, and no one could do better than to study them and try to equal them in clearness and general excellence. According to the writer's observation no photo-micrographs of histological objects have ever exceeded those made by Woodward, and most of them are vastly inferior. It is gratifying to state, however, that at the present time (1896) many original papers are partly or wholly illustrated by photo-micrographs, and no country has produced works with photo-micrographic illustrations superior to those in "Wilson's Atlas of Fertilization and Karyokinesis" and "Starr's Atlas of Nerve Cells," issued by the Columbia University Press.

briefly recapitulate, photography is apparently somewhat indebted to microscopy for the first fleeting pictures of Wedgwood and Davy [1802], the first methods of producing permanent paper prints [Reede, 1837-1839], the first offering of prints for sale, the first plates engraved after photographs for the purpose of book illustration [Donné & Foucault, 1845], the photographic use of collodion [Archer & Diamond, 1851], and finally, wholly indebted for the origin of the gelatino-bromide process, greatest achievement of them all [Dr. R. L. Maddox, 1871]. See further for the history of Photo-micrography, Neuhauss, also Bousfield.

As the difficulties of photo-micrography are so much greater than of ordinary photography, the advice is almost universal that no one should try to learn photography and photo-micrography at the same time, but that one should learn the processes of photography by making portraits, landscapes, copying drawings, etc., and then when the principles are learned one can undertake the more difficult problem of photo-micrography with some hope of success.

The advice of Sternberg is so pertinent and judicious that it is reproduced "Those who have had no experience in making photo-micrographs are apt to expect too much and to underestimate the technical difficulties. Objects which under the microscope give a beautiful picture, which we desire to reproduce by photography, may be entirely unsuited for the purpose. In photographing with high powers it is necessary that the objects to be photographed be in a single plane and not crowded together and overlying each other. For this reason photographing bacteria in sections presents special difficulties and satisfactory results can only be obtained when the sections are extremely thin and the bacteria well stained. Even with the best preparations of this kind much care must be taken in selecting a field for photography. It must be remembered that the expert microscopist, in examining a section with high powers, has his finger on the fine adjustment screw and focuses up and down to bring different planes into view. He is in the habit of fixing his attention on the part of the field which is in focus and discarding the rest. But in a photograph the part of the field not in focus appears in a prominent way which mars the beauty of the picture."

APPARATUS FOR PHOTO-MICROGRAPHY.

§ 324. **Camera.**—For the best results with the least expenditure of time one of the cameras especially designed for photo-micrography is desirable but is not by any means necessary for doing good work. An ordinary photographic camera, especially the kind known as a copying camera, will enable one to get good results, but the trouble is increased, and the difficulties are so great at best, that one would do well to avoid as many as possible and have as good an outfit as can be afforded.

The first thing to do is to test the camera for the coincidence of the plane occupied by the sensitive plate and the ground glass or focusing screen. Cameras even from the best makers are not always correctly adjusted. By using a straight edge of some kind, one can measure the distance from the inside or ground side of the focusing screen to the surface of the frame. This should be done all around to see if the focusing screen is equally distant at all points from the surface of the frame. If it is not it should be made so. When the focusing screen has been examined, an old plate, but one that is perfectly flat, should be put into the plate holder and the slide pulled out and the distance from the surface of the plate holder determined exactly as for the focusing screen. If the distance is not the same, the position of the focusing screen must be changed to correspond with that of the glass in the plate holder, for unless the sensitive surface occupies exactly the position of the focusing screen the picture will not be sharp no matter how accurately one may focus. Indeed, so necessary is the coincidence of the plane of the focusing screen and sensitive surface that some photo-micrographers put the focusing screen in the plate holder, focus the image and then put the sensitive plate in the holder and make the exposure (Cox). This would be possible with the older forms of plate holders, but not with the double plate holders mostly used at the present day.

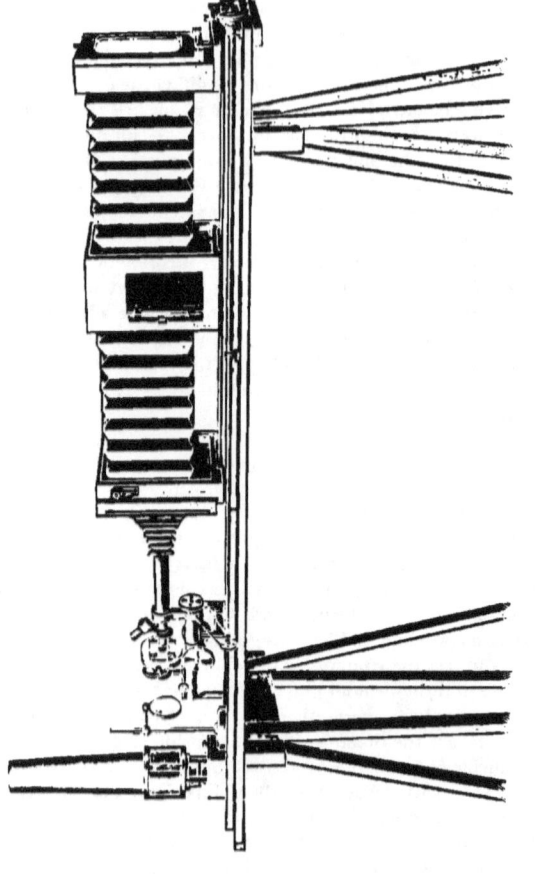

Fig. 143.

FIG. 143. *Walmsley's Improved Photo-Micrographic Camera. In this figure is shown an excellent form of photo-micrographic camera for use with a horizontal microscope. It has the advantage of the possibility of a very long bellows or a shorter one as the need of the special case demands. It is arranged for photo-micrographic work with the microscope or a wide angled objective or for copying and slightly enlarging or diminishing, drawing, etc., with an ordinary photographic objective. It is also arranged for making lantern slides. A very simple arrangement has been adopted for focusing when the bellows is pulled out so far that one cannot reach the fine adjustment, and it works with great smoothness. "A groove is turned in the periphery of the fine adjustment screw, around which a small cord is passed, and carried through a succession of screw-eyes on either side of the baseboard to the rear, where a couple of small leaden weights are attached to its ends, thus keeping the cord taut. A very slight pull on either side, whilst the eye is fixed upon the image on the screen, suffices to adjust the focus with the utmost exactness. If preferred, a rod running the entire length of the camera-bed, and terminating at the rear end with a milled head, whilst on the other is a grooved pulley, carrying a cord or belt which also passes around the groove in the milled head of the fine adjustment, may be substituted for the cord and weights. This arrangement [shown in the figure] is more costly, and probably no better in actual service than that with the cord and weights."* (From Mr. Walmsley.)

§ 325. **Work Room.**—It is almost self-evident that the camera must be in some place free from vibration. Frequently a basement room where the camera table may rest directly on the ground or on a pier is an excellent situation. Such a situation is almost necessary for the best work with high powers. For those living in cities, a time must also be chosen when there are no heavy vehicles moving in the streets. For less difficult work an ordinary room in a quiet part of the house or laboratory building will suffice.

§ 326. **Arrangement and Position of the Camera and the Microscope.**—For the greater number of photo-micrographs, a horizontal camera and microscope are to be preferred as one may then vary the length of the bellows at will and still preserve steadiness (Fig. 143). For some specimens, however, it is necessary to keep the microscope in a vertical position, hence to photograph, the camera must also be vertical. Very excellent arrangements were perfected long ago, especially by the French. (See Moitessier).

Vertical photo-micrographic cameras are now very commonly made, and by some firms only vertical cameras are produced. They are exceedingly convenient, and do not require so great a disarrangement of the microscope to make the picture. Van Heurck advises their use, then whenever a structure is shown with especial excellence it is photographed immediately. The variation in size of the picture is obtained by the objective and the projection ocular rather than by length of bellows (see below). (Fig. 144). It must not be forgotten, however, that penetration varies inversely as the *square* of the power, and only inversely as the numerical aperture (§ 29), consequently there is a real advantage in using a low power of great aperture and a long bellows rather than an objective of higher power with a shorter bellows. (See Carpenter-Dallinger, pp. 318-319).

§ 327. **Microscope.**—For convenience and rapidity of work a microscope with mechanical stage is very desirable. It is also an advantage to have a tube of large diameter so that the field will not be too greatly restricted. In some microscopes the tube is removable almost to the nose-piece to avoid interfering with the size of the image. The substage condenser should be movable on a rack and pinion. The microscope should have a flexible pillar for work in a horizontal position. While it is desirable in all cases to have the best and most convenient apparatus that is made, it is not by any means necessary for the production of excellent work. A simple stand with flexible pillar and good fine adjustment will answer.

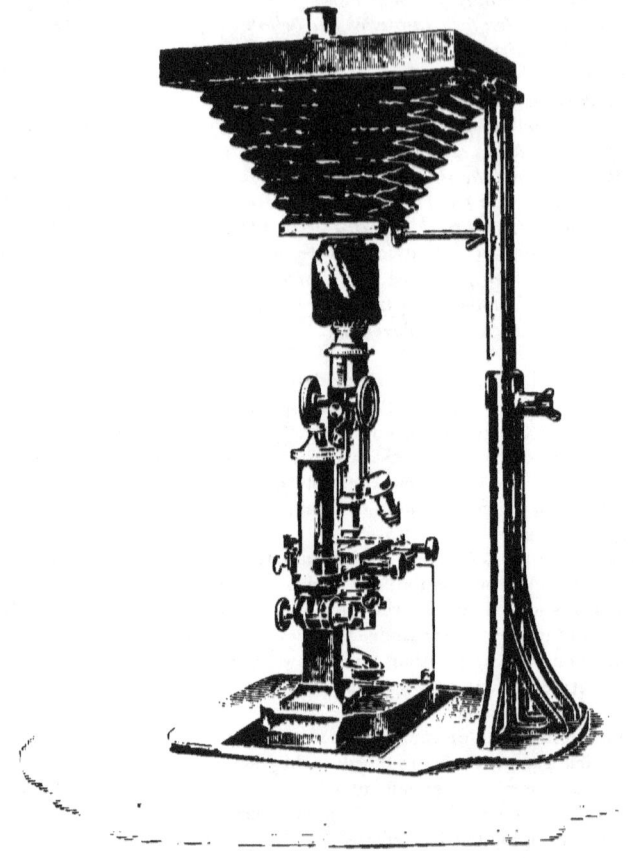

FIG. 144.

FIG. 144. *Leitz' Photo-Micrographic Camera for Vertical Microscope (from Leitz' American House, New York). The camera may be raised or lowered to adapt it to various stands. The microscope, as shown, rests on the base of the camera support, thus aiding steadiness, and the simultaneous vibration of the entire apparatus, if any should occur.*

With this instrument the focusing can be done with the hand on the fine adjustment, as in ordinary microscopical study.

§ 328. **Objectives and Oculars for Photo-Micrography.**—The belief is almost universal that the apochromatic objectives are most satisfactory for photography. They are employed for this purpose with a special projection ocular. Two very low powers, one of 35 and one of 70 millimeters equivalent focus are used without any ocular (Fig. 149). Some of the best work that has ever been done, however, was done with achromatic objectives (work of Woodward and others). One need not desist from undertaking photo micrography if he has good achromatic objectives. From a somewhat extended series of experiments with the objectives of many makers the good modern achromatic objectives were found to give excellent results when used without an ocular. Most of them also gave good results with projection oculars, although it must be said that the best results were obtained with the apochromatic objectives and projection oculars. It does not seem to require so much skill to get good results with the apochromatics as with the achromatic objectives. The majority of photo-micrographers do not use the Huygenian oculars in photography, although excellent results have been obtained with them. An amplifier is sometimes used in place of an ocular. Considerable experience is necessary in getting the proper mutual position of objective and amplifier. The introduction of oculars especially designed for projection, has led to the discarding of ordinary oculars and of amplifiers. However the projection oculars of Zeiss restrict the field very greatly, hence the necessity of using the objective alone for large specimens.*

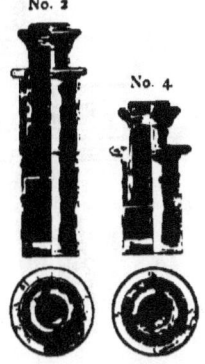

FIG. 145. *Projection Oculars with section removed to show the construction. Below are shown the upper end with graduated circle to indicate the amount of rotation found necessary to focus the diaphragm on the screen. No. 2, No. 4. The numbers indicate the amount the ocular magnifies the image formed by the objective as with the compensation oculars. (Zeiss' Catalog, No. 30).*

§ 329. **Difference of Visual and Actinic Foci.** Formerly there was much difficulty experienced in photo-micrographing on account of the difference in actinic and visual foci. Modern objectives are less faulty in this respect and the apochromatics are practically free from it. Since the introduction of orthochromatic or isochromatic plates and, in many cases the use of colored screens, but little trouble has

---

*The comparative study both with projection oculars, and without an ocular were made with the achromatic objectives 25 mm. (1 inch), 18 mm. (¾ inch), 5 mm. ¹ to ⅛ inch) and 2 mm. $\frac{1}{12}$ (inch) homogeneous immersion of the Bausch & Lomb Optical Co.; Gundlach Optical Co.; Leitz; Reichert; Winkel and Zeiss. Good results were obtained with all of these objectives both with and without projection oculars. The objectives of Spencer and Smith, especially constructed for photo-micrography, are highly spoken of by Piffard (Jour. Roy. Micr. Soc., 1892-1893). The writer has not had the opportunity of comparing these with those mentioned. Piffard spoke highly of the work of Wales also. From the known excellence of the work of these opticians one would expect good results.

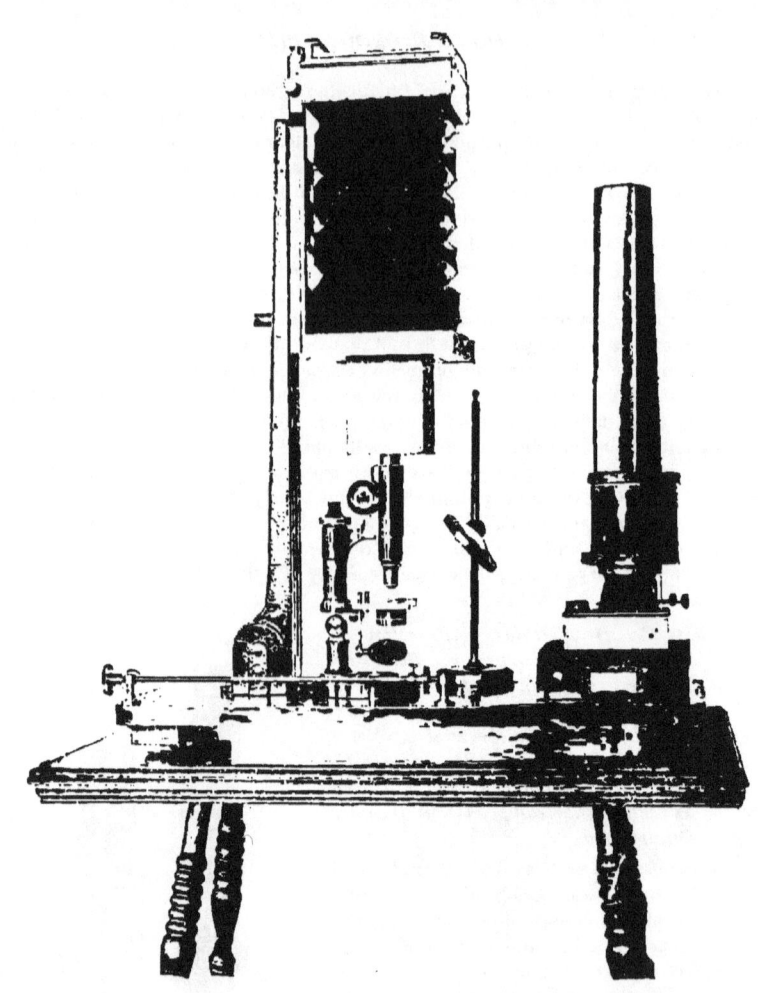

FIG. 148. *Walmsley's Autograph Photo-Micrographic Camera in a vertical position. Compare Fig. 147 (From the Proc. Amer. Micr. Soc. Vol. XVII, 1895).*

screen. They should be carefully tested to see if there is coincidence in position of the focusing screen and the sensitive film as described in (§ 324.

§ 330. **Apparatus for Lighting.**—For low power work (35 mm. and longer focus) and for large objects some form of bull's eye condenser is desirable although fairly good work may be done with diffused light or lamp-light reflected by a mirror. If a bull's eye is used it should be as nearly achromatic as possible. The engraving glass shown in Fig. 155 answers very well for large objects. For smaller objects a Steinheil lens combination gives a more brilliant light and one also more nearly achromatic. For high power work all are agreed that nothing will take the place of an achromatic condenser. This may be simply an achromatic condenser, but preferably it should be an *apochromatic* condenser. Whatever the form of the condenser it should possess diaphragms so that the aperture of the condenser may be varied depending upon the aperture of the objective. For a long time, objectives have been used as achromatic condensers, and they are very satisfactory, although less convenient than a special condenser whose aperture is great enough for the highest powers and capable of being reduced by means of diaphragms to the capacity of the lower objectives. It should also be capable of accurate centering.

§ 331. **Light Filters or Color Screens.**—These are solutions or suitably stained collodion or gelatin films placed between the source of illumination and the object. It does not make much difference where the color screen is placed provided no light reaches the object which has not passed through the filter. The purpose of the color screen or filter is to take out the excessive number of blue and violet rays so that the more slowly acting red, yellow and green may have time to produce the appropriate chemical changes in the sensitive plate. This action of the longer waves (see under spectroscope § 179-198), is greatly aided by the isochromatic or orthochromatic plates which are especially treated so that they will be sensitive to the longer waves as well as to the shorter, blue and violet waves. This is why it is so necessary in manipulating the plates to avoid exposing them to any light whatsoever in putting them into the plate holder, developing, etc.

The color screen used by Dr. Leaming in preparing the negatives for the plates in Wilson's and Starr's atlases was made by staining a lantern slide plate from which all the silver salts had been removed, with an alcoholic solution *of tropaeolin* and then after drying, Canada balsam and a coverglass were applied. Others have recommended collodion stained with aurantia. The purpose of these is to filter out the greater number of the blue and violet rays and then increase the time of exposure from 2 to 5 times, depending on the thickness of the color screen. Color screens and color cells are furnished by various makers. None of them answers for all preparations and for some preparations they are unnecessary.

§ 332. **Objects Suitable for Photo-micrographs**—While almost any large object may be photographed well with the ordinary camera and photographic objective, only a small part of the objects mounted for microscopic study can be photo-micrographed satisfactorily. Many objects that give beautiful and satisfactory images when looking into the microscope and constantly focusing with the fine adjustment, appear almost without detail on the screen of the photo micrographic camera and in the photo-micrograph.

If one examines a series of photo-micrographs the chances are that the greater number will be of diatoms, plant sections or preparations of insects. That is, they are of objects having sharp details and definite outlines, so that contrast and definiteness may be readily obtained. Stained microbes also furnish favorable objects when mounted as cover glass preparations.

Preparations in animal histology must approximate as nearly as possible to the conditions more easily obtained with vegetable preparations. That is, they must be made so thin and be so prepared that the cell outlines will have something of the definiteness of vegetable tissue. It is useless to expect to get a clear photograph of a section in which the details are seen with difficulty when studying it under the microscope in the ordinary way.

Many sections which are unsatisfactory as wholes, may nevertheless have parts in which the structural details show with satisfactory clearness. In such a case the part of the section showing details satisfactorily should be surrounded by a delicate ring by means of a marker (see Figs. 61-66). If one's preparations have been carefully studied and the special points in them thus indicated, they will be found far more valuable both for ordinary demonstration and for photography. The amount of time saved by marking one's specimens can hardly be overestimated. The most satisfactory material for making the rings is shellac colored with an alcoholic solution of one of the anilins, blue or green, then in studying the preparation one can see it even where covered by the ring.

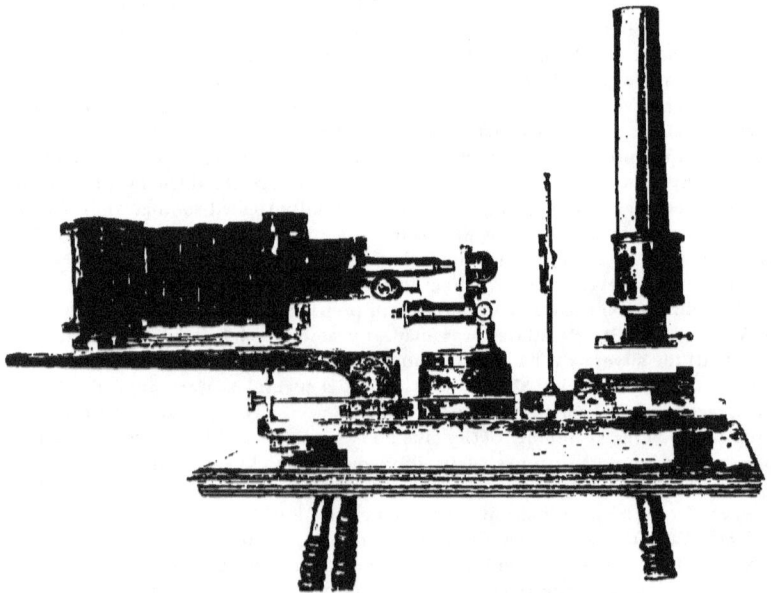

FIG. 147. *Walmsley's Autograph Photo micrographic Camera in a horizontal position. A microscope lamp and bull's-eye condenser are in position. Compare Fig. 148 in (Proc. Amer. Micr. Soc., Vol. XVII, 1895).*

§ 333. **Light.**—The strongest available light is sunlight. That has the defect of not always being available, and of differing greatly in intensity from hour to hour, day to day and season to season. The sun does not shine in the evening when many workers find the only opportunity for work. Following the sunlight, the electric light is the most intense of the available lights. Then comes magnesium, the lime light, the gas-glow or Welsbach light, and lastly, petroleum light. The last is excellent for the majority of low and moderate power work. And even for 2 mm. homogeneous immersion objectives, the time of exposure is not excessive for many specimens (1½ to 3 minutes). This light is also cheapest and most available.

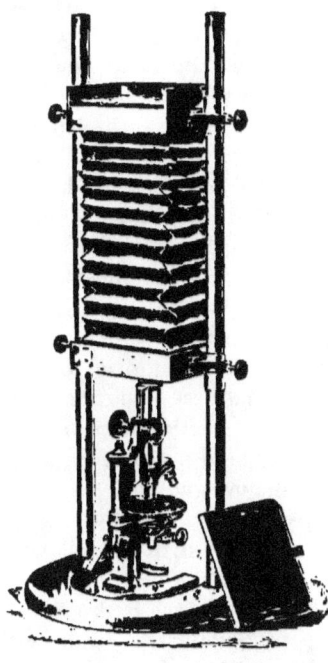

FIG. 148. *Vertical Photo-Micrographic Camera furnished by the Bausch & Lomb Optical Company.* (*From the 15th edition (1896) of their Catalog.*)

EXPERIMENTS IN PHOTO-MICROGRAPHY.

§ 334. The following experiments are introduced to show practically just how one would proceed to make photo-micrographs with various powers, and be reasonably certain of fair success. If one consults prints or the published figures made directly from photo-micrographs it will be seen that, excepting the bacteria, the magnification ranges mostly beween 10 and 150 diameters. The technical difficulties in making good photo-micrographs of animal tissues at a greater magnification are so great that, while they may be used as the basis for figures, they are, in most cases not suitable for direct reproduction.

§ 335. **Photo-Micrographs at a Magnification of 5 to 20 Diameters.**—In the study of embryology and the morphology of small animals or of individual organs like the brain, it is frequently desirable to make pictures of the whole object in its natural setting. These objects and their surroundings are frequently from one to two centimeters in diameter, that is of a size too great to be satisfactorily photographed with microscopic objectives. In common with other observers the writer has found the short focus, wide angled, photographic objectives to give excellent results. It is necessary to have considerable length (1½ to 2 meters) of bellows for this, if a magnification of 10 to 15 diameters is desired.

Put the objective in position in the front of the camera and place the object on some kind of support near the objective. A T-shaped board with a hole of proper size is good. Then with a glass chimney on the lamp, place the bull's eye between the lamp and object (Fig. 157). Put a piece of white paper over the object and mutually arrange lamp and bull's eye till a sharp image of the flame may be seen on the paper covering the object (Fig. 157). Remove the paper from the object and proceed to focus. This is accomplished by sliding the whole camera toward or away from the object, see also § 339. One must also shorten or lengthen the bellows to get the picture of the proper size.

In case the whole specimen is not illuminated the lamp must be turned so that the broad side of the flame is toward the object. The mutual arrangement of lamp and bull's eye must also be such that the brightest part of the flame illuminates the object. If the bull's eye is moved a little toward the lamp, the flame will be sufficiently broadened. It must be remembered, however, that the best and most intense illumination is obtained when the object appears in the image of the flame. When the object is evenly illuminated and the focus is made as perfect as possible, a small diaphragm is used in the objective (i. e., f 32 or 64) and the plate holder with the sensitive plate is put in place of the focusing screen. Some kind of cover is put over the objective, the slide of the plate holder is withdrawn and then the objective is uncovered. With instantaneous, orthochromatic or isochromatic plates the exposure in most cases need not be over 30 to 90 seconds, depending on the object and the diaphragm.

It sometimes occurs that the whole object cannot be satisfactorily lighted. In that case one may use diffused day-light as follows: Elevate the camera so that the object is against the clear sky for a back ground. If any of the earth should form part of the back ground that part of the object would not be sufficiently illuminated. It is best not to use any mirror. The light from the sky will evenly and completely illuminate even the largest object.

Instead of elevating the camera one might use a large reflector covered with very white cloth or paper and set at an angle of 45 degrees. This reflector would then serve for back ground equally with the clear sky.

§ 336. **Focusing Screen for Photo-Micrography.** One cannot expect a picture sharper than the image seen on the focusing screen. Hence the greatest care must be taken in focusing. The general focusing may be done with the unaided eye and on the ground glass, but for the final focusing a clear screen and a focusing glass must be used (Fig. 150).

FIG. 149. *Zeiss' Apochromatic, Projection Objective of 70 mm. equivalent focus, for photo-micography. (Zeiss' Catalog. No. 30). This, and another of 35 mm. focus, are designed for making pictures of moderate magnification. Usually rather large objects are photographed with them. The object may be illuminated in the ordinary way. They are used without an ocular, like a photographic objective. The one of 35 mm. is screwed into the tube of the microscope like an ordinary objective, but the one of 70 mm. here shown, is, by means of a conical adapter, screwed into the ocular end of the tube.*

For illuminating the object, any suitable light may be used, but it is recommended that the light be concentrated by means of a bull's eye or some form of combination like the engraving glass, and that the condenser be so placed that it focuses the light upon the objective, not upon the object. The object is then illuminated with a converging cone of light.

For the clear screen, Mr. Walmsley and others have recommended that a pencil mark or cross be made in the center of the ground glass, and then that a large circular or square cover-glass be put on the ground glass with Canada balsam. To do this, warm the ground glass carefully, add a drop of rather thick balsam to the center on the ground side, then apply the cover and press it down firmly. After the balsam has cooled it may be cleaned off around the cover with xylene or alcohol. The balsam will fill up the inequalities in the glass and being of about the same refractive power will make this part of the glass clear as if it were unground (Fig. 150).

For using the focusing glass first carefully adjust it so that the pencil cross in the center of the ground glass is in the best possible focus. The image when in the best focus must then be in the same plane as

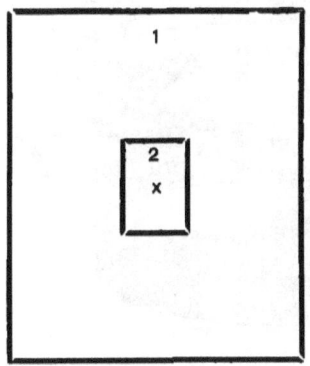

FIG. 150. *Focusing screen with clear center for the final adjustment with a focusing glass like that shown in Fig. 153 or 154.*

1. *The ground surface of the focusing screen; it is translucent but not transparent.*

2. *Central clear part of the screen made by cementing a cover-glass to the ground surface with Canada balsam. In the center is shown the pencil mark to indicate the plane to which the focusing glass should be adjusted.*

the ground side of the focusing screen. If the uncovered part of the focusing screen is too opaque, rub some fine oil on it; only a little should be used. The focusing screen as thus prepared with a clear center, serves both for the general focusing and the finest focusing, and avoids the danger of using a double screen. That is, the fewer the processes the less the liability to error.

FIG. 151. *Perigraphic Objective of about 90 millimeters equivalent focus for making photo-micrographs at a magnification of 2 to 15 diameters. It was found to give the best results when used right end to as in ordinary photography. (From the Gundlach Optical Co.)*

FIG. 152. *Zeiss' Anastigmat Objective of about 85 millimeters equivalent focus for photo-micrographs at a magnification of 2 to 15 diameters. Excellent results were obtained with this, and the other short focused anastigmats. On the whole the results were more satisfactory when the lens was used right end to as in ordinary photography. (From the Bausch and Lomb Optical Co.)*

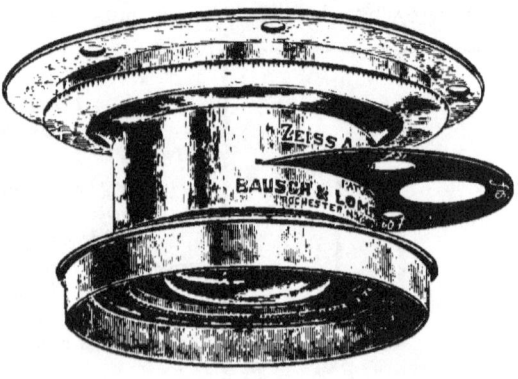

§ 337. **Development of the Negative.**—After the exposure, comes the development; this in photo-micrography requires more judgment than for ordinary photography. The ordinary negative is liable to have too much contrast, but this is rarely the case with photo-micrographs. Any good developer may be used. One can as a rule do no better than to follow the directions accompanying the plates used. The writer's experience has been so satisfactory with Mr. Walmsley's developers that he desires to call attention to them. The developers are easily made, will develop anything that can be developed and one can feel confident that if the negative is not good the fault does not lie with the developer.

The best photo-micrographic negatives made by the writer were made with Cramer's instantaneous isochromatic plates, and the image commenced to appear with Walmsley's developer in 2½ to 3 minutes and the development was completed in 12 to 15 minutes. Excellent negatives have been developed in less time and also when it required half an hour to develop them. The temperature has much to do with the development of correctly exposed negatives, so that no rule can be given. Metol and rodinal developers have also given excellent results and in a shorter time.

If one desires the best possible results it is necessary to avoid the light in developing. Even the ruby light of the dark room should be avoided as much as possible, for the plates used are made purposely sensitive to the longer rays of the spectrum (§ 193). After the negative is developed and washed it should not be taken to the light till it is fixed. Too much light after development and before fixing, injures the clearness of the negative.

FIG. 153. *Focusing Glass.* "*It is achromatic, consisting of a double convex crown lens and a negative meniscus flint lens cemented together.*" *It screws into the brass tube and is thus adjustable, enabling one to focus the pencil mark in the clear area of the focusing screen (Fig. 150) with great accuracy. It also serves to focus the image with ease and accuracy. The eye must not be too close to the upper end of the focusing glass or the field will be restricted. (From the Gundlach Optical Co.)*

§ 338. **Labeling and Care of Negatives.**—The care, printing, etc of negatives is like that of ordinary negatives. It is well to label them.

carefully, however. This label should contain as far as possible, the following information: (1) Kind of plate; (2) Time of exposure; (3) Light used; (4) Objective; (5) Ocular; (6) Magnification; (7) Name of object and number of preparation from which taken; (8) Date of making the negative.

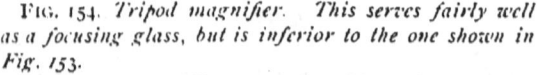

FIG. 154. *Tripod magnifier. This serves fairly well as a focusing glass, but is inferior to the one shown in Fig. 153.*

§ 339. **Low Microscopic Objectives.**—Microscopic objectives of 35 mm. and greater focal length may be attached to the camera directly after the manner of a photographic objective if one has a plate made with the proper screw for the objective. The only difficulty is in focusing the object properly. If one has the special stand made by the Bausch and Lomb Optical Co., or something equivalent so that the object may be moved with a fine screw, the focusing can be well done. Such an arrangement is also very convenient in using the photographic objectives (Figs. 151-152) and the apochromatic objectives of 35 and 70 mm. focus made by Zeiss.

FIG. 155. *Engraving glass to serve as a bull's-eye condenser. (From the Bausch and Lomb Optical Co.).*

§ 340. **Photo-Micrographs of 20 to 50 Diameters.**—For these, low objectives are used (35 mm. to 20 mm. focus). They are attached to the camera as described in § 339 or a microscope is used. If the microscope is used the object is placed on the stage and focused in the usual way. The mirror is preferably removed or swung aside and the lamp and bull's eye mutually arranged to give an image of the flame as described in § 335. If the object is small the achromatic condenser may be used (See § 342) or one of the Steinheil magnifiers. When the light is satisfactory as seen through an ordinary ocular, remove the ocular.

(A) *Photographing without an Ocular.*—After the removal of the ocular put in the end of the tube a lining of black velvet to avoid reflections. Connect the microscope with the camera, making a light tight joint and focus the image on the focusing screen. It will be necessary to focus down considerably to make the image clear. Lengthen or shorten the bellows to make the image of the desired size, then focus with the utmost care. In case the field is too much restricted on account of the tube of the microscope, remove the draw tube. When all is in readiness it is well to wait for three to five minutes and then to see if the image is still sharply focused. If it has got out of focus simply by standing, a sharp picture could not be obtained. If it does not remain in focus, something is faulty. When the image remains sharp after focusing make the exposure. From 15 to 60 seconds will usually be sufficient time with instantaneous plates and the light as described.

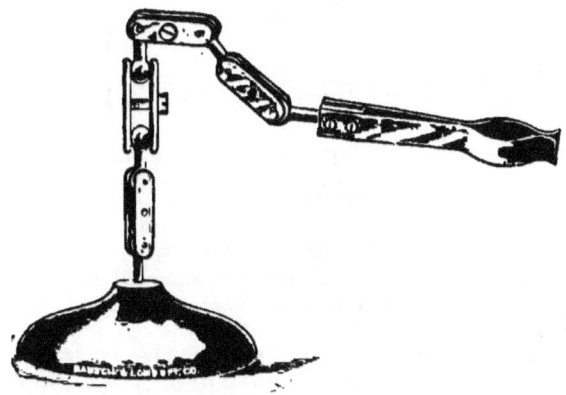

FIG. 156. *Lens holder composed of several links and balls, thus giving the flexibility of a chain and enabling one to turn the lens in any desired direction. Each link has a screw to take up wear, thus insuring permanence. The lens is grasped by two movable pieces which open widely enough to take a tripod or an engraving glass; they also hold with equal facility a very much smaller lens. This is one of the most efficient and satisfactory lens holders ever made. It is especially good for holding a lens for concentrating the light on the mirror in artificial lighting* (From the Bausch & Lomb Optical Co.)

(B) *Photographing with a Projection Ocular.*—If the object is small enough to be included in the field of a projection ocular, one may put that in place of the ordinary ocular. The first step is then to focus the diaphragm of the projection ocular sharply on the focusing screen. Bring the camera up close to the microscope and then screw out the eye-lens of the ocular a short distance. Observe the circle of light on the focusing screen to see if its edges are perfectly sharp. If not, continue to screw out the eye-lens until it is. If it cannot be made sharp by screwing it out reverse the operation. Unless the edges of the light circle, i. e., the diaphragm of the ocular, is sharp, the resulting picture will not be satisfactory. When the diaphragm is sharply focused on the screen, the microscope is focused exactly as though no ocular were present, that is, first with the unaided eye then with the focusing glass; the object should be in focus in the beginning.

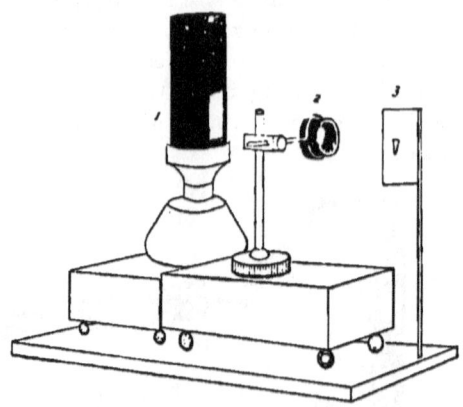

FIG. 157. *Arrangement for Artificial Illumination.*

1. Lamp with metal chimney, easily made by rolling up some ferrotype plate and making a slit-like opening in one side. This opening should be covered by an oblong cover-glass. A glass slide, being of considerable thickness, breaks too easily. The lamp should have a wick about 30 mm. wide, so that the thickness of the flame, if taken edgewise, will give an intense light. A wide flame also enables one to get a larger image of the flame, and thus illuminate a larger object than as though a small flame were used.

2. Bull's-eye condenser on a separate stand. The engraving glass shown in Fig. 155, or the tripod magnifier (Fig. 154) answers fairly. The Steinheil lenses are still better.

3. Screen showing image of the flame inverted.

The lamp and bull's-eye stand are on blocks with screw-eyes as leveling screws.

The exposure is also made in the same way, although one must have regard to the greater magnification produced by the projection ocular and increase the time accordingly; thus when the × 4 ocular is used, the time should be at least doubled over that necessary when no ocular is employed.

Zeiss recommends that when the bellows have sufficient length the lower projection oculars be used, but with a short bellows the higher ones. It is also sometimes desirable to limit the size of the field by putting a smaller diaphragm over the eye lens. This would aid in making the field uniformly sharp.

§ 341. **Photo-Micrographs at a Magnification of 100 to 150 Diameters.**—For this, the simple arrangements given in the preceding section will answer, but the objectives must be of shorter focus, 8 to 3 mm. It is better, however, to use an achromatic condenser instead of the engraving glass or the Steinheil lens.

§ 342. **Lighting for Photo-Micrography with Moderate and High Powers.**—(100 to 2,500 diameters). No matter how good one's apparatus, successful photo-micrographs cannot be made unless the object to be photographed is properly illuminated. The beginner can do nothing better than to go over with the greatest care the directions for centering the condenser, for centering the source of illumination, and the discussion of the proper cone of light and lighting the whole field, as given on pp. 39 to 46. Then for each picture the photographer must take the necessary pains to light the object properly. An achromatic condenser is almost a necessity (§ 76). Whether a color-screen should be used depends upon judgment and that can be attained only by experience. In the beginning one may try without a screen, and with different screens and compare results.

A plan used by many skillful workers is to light the object and the field around it well and then to place a metal diaphragm of the proper size in the camera very close to the plate holder. This will insure a clean, sharp margin to the picture. Of course this metal diaphragm must be removed while focusing the diaphragm of the projection ocular, as the diaphragm opening would be smaller than the image of the ocular diaphragm.

If the young photo-micrographer will be careful to select for his first trials objects of which really good photo-micrographs have already been made, and then persists with each one until fairly good results are attained, his progress will be far more rapid than as if poor pictures of many different things were made. He should, of course, begin with low magnifications.

§ 343. **Adjusting the Objective for Cover-Glass.**—After the object is properly lighted, the objective, if adjustable, must be corrected for the thickness of cover. If one knows the exact thickness of the cover and the objective is marked for different thickness, it is easy to get the adjustment approximately correct mechanically, then the fine or final correction depends on the skill and judgment of the worker. It is to be noted too that if the objective is to be used without a projection ocular the tube length is practically extended to the focusing screen and as the effect of lengthening the tube is the same as thickening the cover-glass, the adjusting collar must be turned to a higher number than the actual thickness of the cover calls for (see § 96).

§ 344. Photographing Without an Ocular.—Proceed exactly as described for the lower power, but if the objective is adjustable make the proper adjustment for the increased tube-length.

§ 345. Photographing with a Projection Ocular.—Proceed as described in § 343, only in this case the objective is not to be adjusted for the extra length of bellows. If it is corrected for the ordinary ocular, the projection ocular then projects this correct image upon the focusing screen.

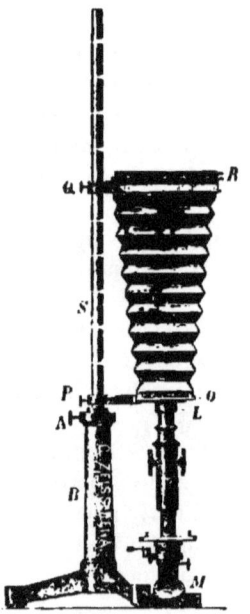

FIG. 158. *Zeiss' Vertical Photo-micrographic Camera. A. Set screw holding the rod (S) in any desired position. P, Q. Set screws by which the bellows are held in place. B. Stand with tripod base in which the supporting rod (S) is held. This rod is now graduated in centimeters and is a ready means of determining the length of the camera. M. Mirror of the microscope. L. The sleeve serving to make a light tight connection between the camera and microscope. O. The lower end of the camera. R. The upper end of the camera where the focusing screen and plate holder are situated. (From Zeiss' Photomicrographic Catalog).*

§ 346. Determination of the Magnification of the Photo-Micrograph.—After a successful negative has been made, it is desirable and important to know the magnification. This is easily determined by removing the object and putting in its place a stage micrometer. If now the distance between two or more of the lines of the micrometer is obtained with dividers and the distance measured on one of the steel rules the magnification is obtained by dividing the size of the image by the known size of the object (§ 146). If now the length of the bel-

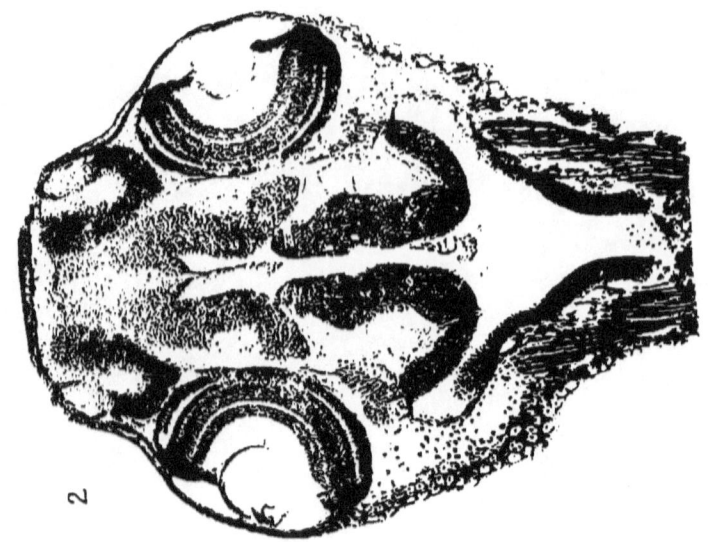

Fig. 160.

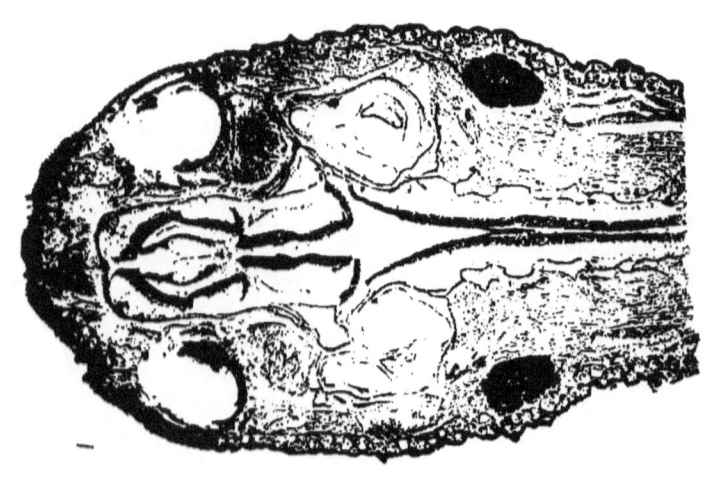

Fig. 161.

FIGS. 160-161. *Fine tint, half-tone reproductions of photo-micrographs of sections made by Mrs. Gage, to show the possibilities of photo-micrography with photographic objectives and with low microscopic objectives without a projection ocular.*

1. Frontal section of the head of a large red Diemyctylus viridescens (red newt) at the level of the portae of the brain, magnified 10 diameters. Negative made with a Gundlach perigraphic objective of about 90 mm. equivalent focus.

2. Frontal section of a larval Diemyctylus about 10 millimeters in length. Negative made with a Winkel objective of 22 millimeters equivalent focus ; no ocular. Magnified 50 diameters. By the permission of Mrs. Susanna Phelps Gage, from the Wilder Quarter Century Book.

lows from the tube of the microscope is noted, say on a record table like that in section 348, one can get a very close approximation to the power at some other time by using the same optical combination and length of bellows.

FIG. 159. *Rack for drying negatives. (From the Rochester Optical Co.)*

§ 347. **Photo-Micrographs at a Magnification of 500 to 2000 Diameters.**— For this the homogenous immersion objectives should be employed, and as it would require a long bellows to get the higher magnification with the objective alone, it is best to use the projection oculars.

For this work the directions given in § 342 must be followed with great exactness. The edge of the lamp flame will be sufficient to fill the field in most cases. With many objects the time required with good lamp light is not excessive ; viz., $1\frac{1}{2}$ to 3 minutes. The reason of this is that while the illumination diminishes directly as the square of the magnification, it increases with the increase in numerical aperture, so that the illuminating power of the homogeneous immersion is great in spite of the great magnification (§ 31).

For work with high powers a stronger light than the petroleum lamp is employed by those doing considerable photo-micrography. Very good work may be done, however, with the petroleum lamp.

It may be well to recall the statement made in the beginning, that the specimen to be photographed must be of especial excellence for all powers. No one will doubt the truth of the statement who undertakes to make photo-micrographs at a magnification of 500 to 2000 diameters.

§ 348. RECORD TABLE FOR PHOTO-MICROGRAPHY.

Camera and Objective.	Ocular.	Magnification and Length of Bellows.	Condenser.	Light, Hour, Date.	Exposure and Remarks.	Object.	Stain.	Color Screen.	Plates.	Developer.	Remarks.	Negative.	Photographer.
Vertical (Zeiss').	× 4	80	Zeiss' 1 N. A.	Daylight.	35 sec.	Silvered Hepatic Ligt.	Silver	None.	Cramer's Inst. Iso-chromatic.	Rodinal		Negative	Stotsen-burg.
Zeiss' 8 mm. Apochromatic	Projection	Bellows, 50 cm.	Achromatic.	5 P. M., May 10, '98.	Correct.	Necturus.	and Hema-toxylin.			1-30	Good.		

PHOTOGRAPHING NATURAL HISTORY SPECIMENS WITH A VERTICAL CAMERA.*

§ 349. For most natural history specimens it is inconvenient, and for many impossible, to use a horizontal camera and to raise the objects in a vertical position. In order to have the objects horizontal either a mirror must be used or preferably the camera itself may be so arranged that it may be put in a vertical position.

For the last twenty years such a camera has been in use in the Anatomical Department of Cornell University for photographing all kinds of specimens; among these, fresh brains, and hardened brains have been photographed without the slightest injury to them. Furthermore, as many specimens are so delicate that they will not support their own weight, they may be photographed under alcohol or water with a vertical camera and the result will be satisfactory as a photograph and harmless to the specimen.

A great field is also open for obtaining life-like portraits of water animals. Freshly killed or etherized animals are put into a vessel of water with a contrasting back ground and arranged as desired then photographed. The fins have something of their natural appearance and the gills of branchiate salamanders float out in the water in a natural way. In case the fish tends to float in the water a little mercury injected into the abdomen or intestine will serve as ballast.

The photographs obtainable in water are almost if not quite as sharp as those made in air. Even the corrugations on the scales of such fishes as the sucker (Catostomus teres) show with great clearness. Indeed so good are the results that excellent fine tint, half tone plates may be produced from the pictures thus made, also excellent photogravures. In those cases, as in anatomical preparations, where the photograph rarely answers the requirements of a scientific figure, still a photograph serves as a most admirable basis for a scientific figure. The photograph is made of the desired size and all the parts are in correct proportion and in the correct relative position. From this photographic picture may be traced all the outlines upon the drawing paper, and the artist can devote his whole time and energy to giving the proper expression without the tedious labor of making measurements.

" While the use of photography for outlines as bases for figures di-

*Papers on this subject were given by the writer at the meeting of the American Association for the Advancement of Science in 1879, and at the meeting of the Society of Naturalists of the eastern United States in 1883; and in *Science*, Vol III, pp. 443, 444.

minishes the labor of the artist about one-half it increases that of the preparator; and herein lies one of its chief merits. The photographs being exact images of the preparations, the tendency will be to make them with greater care and delicacy, and the result will be less imagination and more reality in published scientific figures; and the objects prepared with such care will be preserved for future reference."

" In the use of photography for figures several considerations arise: $1°$. The avoidance of distortion; $2°$. The adjustment of the camera to obtain an image of the desired size; $3°$. Focusing; $4°$. Lighting and centering the object; $5°$. Obtaining outlines for tracing upon the drawing-paper."

" $1°$. While the camera delineates rapidly, the image is liable to distortion. I believe opticians are agreed, that, in order to obtain correct photographic images, the objective must be properly made, and the plane of the object must be parallel to the plane of the ground glass. Furthermore, as most of the objects in natural history have not plane surfaces, but are situated in several planes at different levels, the whole object may be made distinct by using in the objective a diaphragm with a small opening."

" $2°$. By placing the camera on a long table, and a scale of some kind against the wall, the exact position of the ground glass for various sizes may be determined once for all. These positions are noted in some way (on the brass guide, 3, in the apparatus here figured). Whenever it is desired to photograph an object, natural size, for example, the ground glass is fixed in the proper position indicated on the brass guide (Fig. 162, 3). Then as the relative position of the objective and the ground glass must not be varied, it is necessary, in focusing, to move the camera toward or away from the object, or the reverse. To do this, the camera is fastened to a board which moves in a frame by means of a screw Fig. 162, 7. Whenever the camera is to be moved considerably,—as to a position for twice natural size from one giving an image of half natural size,—the position of the camera on the board is changed by loosening the two thumb-screws clamping it to the movable board (Fig. 162, 5, 6). The approximate position for the various sizes being once determined and noted, it is but a moment's work to set the camera for any enlargement or reduction within its range."

$3°$. The object is placed on a horizontal support, and so arranged that the lighting will give prominence to the parts to be especially emphasized. For a contrasting background, black velveteen for light, and white paper for dark, objects, have been found excellent.

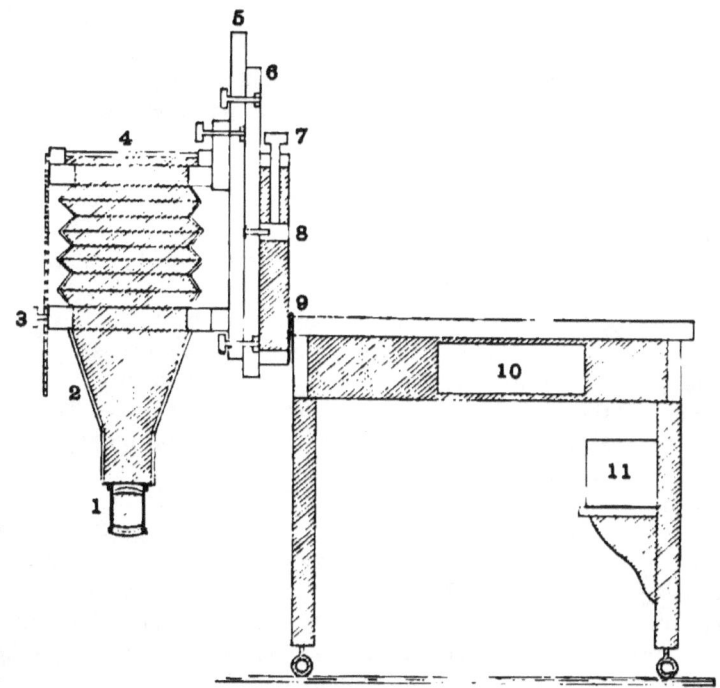

FIG. 162.

Arrangement of a vertical or horizontal camera for making photographs of natural history objects—sectional view.

1. The photographic objective. It should be of the best quality and rectilinear so that there may be no distortion.
2. Cone to increase the length of the camera and avoid shadows.
3. Graduated rod to support the front of the camera and hold it rigid, and also to serve as guide to the various magnifications and reductions most commonly desired.
4. Ground glass. It is an advantage to have a clear space in the middle for accurate focusing, as for the photo-micrographic camera (Fig. 153).
5. Camera bed fastened to the sliding focusing board, 6.
6. Focusing board to which the camera is clamped.
7. Focusing screw.
8. Solid piece connecting the focusing board and focusing screw.
9. Hinge on which the camera swings.
10. Drawer in which are kept the objective and other accessories.
11. Box of sand or other heavy material to serve as ballast.
The large screw eyes in the legs of the table serve as leveling screws.

4°. If the photographic prints are to be used solely for outlines, the well-known blue prints so much used in engineering and architecture may be made. If, however, light and shade and fine details are to be brought out with great distinctness, either an aristotype, platinotype or a bromide print is preferable. In whatever way the print is made, it is blacked on the back with soft lead-pencil, put over the drawing-paper, and the outlines traced.

OUTLINES OBTAINED DIRECTLY BY MEANS OF THE CAMERA.

§ 350. While it is desirable to make photographs of objects in many cases, this may frequently be avoided and a tracing made directly from the camera. The object is arranged as for a photograph and well lighted. Then the ground glass is changed for a plain glass and the tracing paper is put over the glass. If the head and screen are covered in some way the image may be seen and traced very easily. This will give a reversed image, but that is easily remedied by turning the paper over in making the tracing on the drawing paper.

Frequently also one wishes to enlarge or reduce a drawing or outline already made. If the figure or outline is placed in a good light the image may be made of any desired size and either photographed or traced as just described. Tracings of any desired size may also be obtained of negatives, but for this one must employ the method of lighting described in § 335, where the clear sky is taken for a background and illuminant. Of course artificial light may also be used, but it is less satisfactory unless one has abundant facilities.

An excellent method of making large diagrams is to use a photographic objective and project the image into a dark room, something after the manner of a stereopticon, then one can trace the image directly on the material on which the diagram is to be drawn.

§ 351. **Prints of Photo-Micrographs and Mechanical Printing.** After one's negatives are made, prints from them may be obtained from a photographer, but from the author's experience, unless the photographer is familiar with the kind of printing necessary to bring out the features most desired, the results will not be very satisfactory. It is better to make one's own prints, and this can be very easily done with the excellent aristotype and platinotype paper on the market. It may also be well done with the bromide paper. The last has the advantage of being capable of furnishing prints by lamp as well as daylight.

For mechanical prints, half tones and photogravures, one should get first as good a negative as possible. And for photogravures the so-called stripping plate should be used so that the picture will not be reversed. For half tones the print should have a good deal of contrast and some pure white. With good printing the fine half tone work on the larger natural history specimens is capable of giving very satisfactory results as may be seen by observing Fig. 160 and 161.

For the methods of Photo-Micrography and Photography in general, consult the works mentioned in the bibliography. Especial attention is also called to the papers of Woodward "On an improved method of photographing histological preparations by sunlight, 1871, and on the Application of Photography to Micrometry."

To the papers and discussions of A. C. Mercer in the various volumes of the Proceedings of the American Microscopial Society.

To the papers and discussions on photo-micrography by Piffard ; Journal of the Royal Micr. Soc. 1892, and 1893, also in the Amer. Jour. Med. Sciences 1893. The paper by Dr. G. A. Piersol,—"Some experiences in Photo-Micrography" in the American Annual of Photography for 1890 will also be found very instructive and helpful. In every volume, and freqeuntly in every number, of the microscopical journals of this and foreign countries one may find articles, notes or references to work in photo-micrography. Excellent papers are also frequently found in the photographic journals and annuals.

Thomas J. Wray.—Photo-Micrography by use of ordinary objectives, practically considered with specimens of work. Proc. Amer. Micr. Soc. Vol. XVIII, (1896). The specimens in connection with this paper impressed one that simple apparatus was no bar to success.

Sunlight and Heliostat.—In case one wishes to utilize sunlight for Photo-Micrography it would be of great advantage to consult the papers of Woodward mentioned above and also the paper of Mercer in the Jour. Roy. Micr. Soc., 1892, pp. 305 to 318. For work with sunlight, a heliostat is almost a necessity. An excellent and inexpensive form was devised by Dr. L. Deck and described and figured in the Proc. Amer. Micr. Soc., 1891, pp. 49-50. A good form of heliostat is also figured and described in Van Heurck, pp. 265-266.

Sunlight is not so easily managed as some form of artificial light, even when one possesses a heliostat.

For samples of photo-micrographic work, see the magnificent photo prints exhibited by Dr. Woodward at the Centennial Celebration at Philadelphia in 1876, the works of Mason, Sternberg, Piersol, Wilson and Starr; Carpenter-Dallinger, Pringle, Bousfield Nehauss, etc., besides the great embryological and morphological journals where photo-micrographic plates are frequently published. Catalogs of photo-micrographic apparatus, in some cases, have admirable specimens of photo-micrographs.

APPENDIX.

Abbe's Test-Plate; The Apertometer; Experimental Determination of the Equivalent Focus of the Objective and of the Ocular; Testing Homogeneous Fluid; Preparation of Diagrams; Drawings for Photo-Engraving.

§ 352. For the sake of those who may have opportunity to use Abbe's Test-Plate the following directions from Zeiss are appended:

"CARL ZEISS, JENA. ON THE METHOD OF USING ABBE'S TEST-PLATE."

"This test-plate is intended for the examination of objectives with reference to their corrections for spherical and chromatic aberration and for estimating the thickness of the cover glass for which the spherical aberration is best corrected."

"The test-plate consists of a series of cover-glasses ranging in thickness from 0.09 mm. to 0.24 mm., silvered on the under surface and cemented side by side on a slide. The thickness of each is written on the silver film. Groups of parallel lines are cut through the film and these are so coarsely ruled that they are easily resolved by the lowest powers, yet from the extreme thinness of the silver they form a very delicate test for objectives of even the highest power and widest aperture. To examine an objective of large aperture the plates are to be focused in succession observing each time the quality of the image in the center of the field and the variation produced by using alternately central and very oblique illumination. When the objective is perfectly corrected for spherical aberration for the particular thickness of cover-glass under examination, the contour of the lines in the center of the field will be perfectly sharp by oblique illumination without any nebulous doubling or indistinctness of the minute irregularities of the edges. If after exactly adjusting the objective for oblique light central illumination is used no alteration of the adjustment should be necessary to show the contours with equal sharpness."

"If an objective fulfills these conditions with any one of the plates it is free from spherical aberration when used with cover-glasses of that thickness; on the other hand if every plate shows nebulous doubling or an indistinct appearance of the edges of the silver lines, with oblique illumination, or if the objective requires a different adjustment to get equal sharpness with central as with oblique light, then the spherical correction is more or less imperfect."

"Nebulous doubling with oblique illumination indicates overcorrection of the marginal zone, want of the edges without marked nebulosity indicates undercorrection of this zone; an alteration of the adjustment for oblique and central illumination, that is, a difference of plane between the image in the peripheral and central portions of the objective points to an absence of concurrent action of the separate zones, which may be due to either an average under or overcorrection or to irregularity in the convergence of the rays."

"The test of chromatic correction is based on the character of the color bands, which are visible by oblique illumination. With good correction the edges of the silver lines in the center of the field should show but narrow color bands in the complementary colors of the secondary spectrum, namely on one side yellow-green to apple-green on the other violet to rose. The more perfect the correction of the spherical aberration the clearer this color band appears."

"To obtain obliquity of illumination extending to the marginal zone of the objective and a rapid interchange from oblique to central light Abbe's Illuminating apparatus is very efficient, as it is only necessary to move the diaphragm in use nearer to or further from the axis by the rack and pinion provided for the purpose. For the examination of immersion objectives, whose aperture as a rule is greater than 180° in air and those homogeneous-immersion objectives, which considerably exceed this, it will be necessary to bring the under surface of the Test plate into contact with the upper lens of the illuminator by means of a drop of water, glycerin or oil."

"In this case the change from central to oblique light may be easily effected by the ordinary concave mirror but with immersion lenses of large aperture it is impossible to reach the marginal zone by this method, and the best effect has to be searched for after each alteration of the direction of the mirror."

"For the examination of objectives of smaller aperture (less than 40°-50°) we may obtain all the necessary data for the estimation of the spherical and chromatic corrections by placing the concave mirror so far laterally, that its edge is nearly in the line of the optic axis the incident cone of rays then only filling one-half of the aperture of the objective. The sharpness of the contours and the character of the color bands can be easily estimated. Differences in the thickness of the cover-glass within the ordinary limits are scarcely noticeable with such objectives."

"It is of fundamental importance in employing the test as above described to have brilliant illumination and to use an eye-piece of high power."

"When from practice the eye has learnt to recognize the finer differences in the quality of the contour images this method of investigation gives very trustworthy results. Differences in the thickness of cover-glasses of 0.01 or 0.02 mm. can be recognized with objectives of 2 or 3 mm. focus."

"With oblique illumination the light must always be thrown perpendicularly to the direction of the lines."

"The quality of the image outside the axis is not dependent on spherical and chromatic correction in the strict sense of the term. Indistinctness of the contours towards the borders of the field of view arises as a rule, from unequal magnification of the different zones of the objective; color bands in the peripheral portion (with good color correction in the middle) are caused by unequal magnification of the different colored images."

"Imperfections of this kind, improperly called 'curvature of the field,' are shown to a greater or less extent in the best objectives, where the aperture is considerable."

FIG. 164. *Set of lines under one cover-glass in the Abbe Test-Plate.*

DETERMINATION OF THE APERTURE OF OBJECTIVES.

§ 353. **Determination of the Aperture of Objectives with an Apertometer.**—Excellent directions for using the Abbe apertometer may be found in the Jour. Roy. Micr. Soc., 1878, p. 19, and 1880, p. 20; in Dippel, Zimmerman and Czapski. The following directions are but slightly modified from Carpenter-Dallinger, pp 337–338. The Abbe apertometer involves the same principle as that of Tolles, but it is carried out in a simpler manner; it is shown in Fig. 165. As seen by this figure

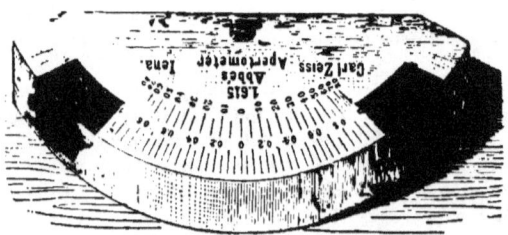

FIG. 165. *Abbe Apertometer.*

it consists of a semi-circular plate of glass. Along the straight edge or chord the glass is beveled at 45°, and near this straight edge is a small, perforated circle, the perforation being in the center of the circle. To use the apertometer the microscope is placed in a vertical position, and the perforated circle is put under the microscope and accurately focused. The circular edge of the apertometer is turned toward a window or plenty of artificial light so that the whole edge is lighted. When the objective is carefully focused on the perforated circle the draw-tube is removed and in its lower end is inserted the special objective which accompanies the apertometer. This objective and the ocular form a low power compound microscope, and with it the back lens of the objective, whose aperture is to be measured, is observed. The draw-tube is inserted and lowered until the back lens of the objective is in focus. "In the image of the back lens will be seen stretched across, as it were, the image of the circular part of the apertometer. It will appear as a bright band, because the light which enters normally at the surface is reflected by the beveled part of the chord in a vertical direction so that in reality a fan of 180° in air is formed. There are two sliding screens seen on either side of the apertometer; they slide on the vertical circular portion of the instrument. The images of these screens can be seen in the image of the bright band. *These screens should now be moved so that their edges just touch the periphery of the back lens.* They act, as it were, as a diaphragm to cut the fan and reduce it, so that its angle just equals the aperture of the objective and no more." "This angle is now determined by the arc of glass between the screens; thus we get an angle in *glass* the exact equivalent of the aperture of the objective. As the numerical apertures of these arcs are engraved on the apertometer they can be read off by inspection. Nevertheless a difficulty is experienced, from the fact that it is not easy to determine the exact point at which the edge of the screen touches the periphery of the back lens, or as we prefer to designate it, *the limit of aperture*, for curious as the expression may appear we have found at times that the back lens of an objective is *larger* than the *aperture* of the objective *requires*. In that case the edges of the screen refuse to touch the periphery."

In determining the aperture of homogeneous immersion objectives the proper immersion fluid should be used as in ordinary observation. So, also, with glycerin or water immersion objectives.

TESTING HOMOGENEOUS IMMERSION LIQUID.

§ 354. In order that one shall realize the full benefit of the homogeneous immersion principle it is necessary that the homogeneous immersion liquid should be truly homogeneous. In order that the ordinary worker may be able to test the liquid used by him, Professor Hamilton L. Smith devised a tester composed of a slip of glass in which was ground accurately a small concavity and another perfectly plain slip to act as cover. (See Proc. Amer. Micr. Soc., 1885, p. 83 . It will be readily seen that this concavity, if filled with air or any liquid of less refractive index than glass, will act as a concave or dispersing lens. If filled with a liquid of greater refractive index than glass, the concavity would act like a convex lens, but if filled with a liquid of the same refractive index as glass, that is, liquid optically homogeneous with glass, then there would be no effect whatever.

In using this tester the liquid is placed in the concavity and the cover put on. This is best applied by sliding it over the glass with the concavity. A small amount of the liquid will run between the two slips, making optical contact on both surfaces. One should be careful not to include air bubbles in the concavity. The surfaces of the glass are carefully wiped so that the image will not be obscured. An adapter with society screw is put on the microscope and the objective is attached to its lower end. In this adapter a slot is cut out of the right width and depth to receive the tester which is just above the objective. As object it is well to employ a stage micrometer and to measure carefully the diameter of the field without the tester, then with the tester far enough inserted to permit of the passage of rays through the glass but not through the concavity, and finally the concavity is brought directly over the back lens of the objective. This can be easily determined by removing the ocular and looking down the tube.

Following Professor Smith's directions it is a good plan to mark in some way the exact position of the tube of the microscope when the micrometer is in focus without the tester, then with the tester pushed in just far enough to allow the light to pass through the plane glass and finally when the light traverses the concavity. The size of the field should be noted also in the three conditions (§ 46-47).

§ 355 The following table indicates the points with a tester prepared by the Gundlach Optical Co., and used with a 16 mm. apochromatic objective of Zeiss, × 4 compensation ocular, achromatic condenser, 1.00 N. A. (Fig. 41) :

TESTER AND LIQUID IN THE CONCAVITY.	SIZE OF THE FIELD.	ELEVATION OF THE TUBE NECESSARY TO RESTORE THE FOCUS.
No tester used	1.825 mm.	Standard position.
Whole thickness of the tester at one end, not over the cavity . .	1.85 mm. .	No change of focus.
Tester with air in the cavity . . .	.6 mm.	Tube raised 6 mm.
Tester with water	1.075 mm.	. . . . 3½ mm.
Tester with 95% alcohol . .	1.15 mm.	. 3 mm.
. . . . kerosene . . .	1.4 mm.	2 mm.
. . . . Gundlach Opt. Co's hom. liquid	1.825 mm.	¹⁄₁₀ mm.
Bausch & Lomb Opt. Co.'s hom. liquid	1.825 mm.	. ¼ mm.
Leitz' hom. liquid	1.825 mm.	. . . . ½ mm.
Zeiss' hom. liquid	1.825 mm.	. . . . ⅓ mm.

It will be seen by glancing at the above table that whenever the liquid in the tester is of lower index than glass, that the concavity with the liquid acts as a concave lens, or in other words like an amplifier (§ 152), and the field is smaller than when no tester is used. It will also be seen that as the liquid in the concavity approaches the glass in refractive index that the field approaches the size when no tester is present. It is also plainly shown by the table that the greater the difference in refractive index of the substance in the concavity and the glass, the more must the tube of the microscope be raised to restore the focus.

If a substance of greater refraction than glass were used in the tester the field would be larger, *i. e.*, the magnification less, and one would have to turn the tube down instead of up to restore the focus. The tester used in these experiments was made by the Gundlach Optical Company of Rochester.

EQUIVALENT FOCUS OF OBJECTIVES AND OCULARS.

§ 356. To work out in proper mathematical form or to ascertain experimentally the equivalent foci of these complex parts with real accuracy would require an amount of knowledge and of apparatus possessed only by an optician or a physicist. The work may be done, however, with sufficient accuracy to supply most of the needs of the working microscopist. The optical law on which the following is based is:—"*The size of object and image varies directly as their distance from the center of the lens.*"

By referring to Figs. 14, 16, 21, it will be seen that this law holds good. When one considers compound lens-systems the problem becomes involved, as the centre of the lens systems is not easily ascertainable hence it is not attempted, and only an approximately accurate result is sought.

§ 357. **Determination of Equivalent Focus of Objectives.**—Look into the upper end of the objective and locate the position of the back lens. Indicate the level in someway on the outside of the objective. This is not the center of the object-ive but serves as an arbitrary approximation. Screw the objective into the tube of the microscope. Remove the field lens from a micrometer ocular, thus making a positive ocular of it (Fig. 21). Pull out the draw-tube until the distance between the ocular micrometer and the back lens is 250 millimeters. Use a stage microm-eter as object and focus carefully. Make the lines of the two micrometers parallel (Fig. 101). Note the number of spaces on the ocular micrometer required to measure one or more spaces on the stage micrometer. Suppose the two microm-eters are ruled in $\frac{1}{10}$ mm. and that it required 10 spaces on the ocular micrometer to enclose 2 spaces on the stage micrometer, evidently then 5 spaces would cover one. The image, A^1B^1 Fig. 21 in this case is five times as long as the object, A, B.—Now if the size of object and image are directly as their distance from the lens it follows that as the size of object is known ($\frac{2}{10}$ mm.), that of the image directly measured ($\frac{10}{10}$ mm.), the distance from the lens to the image also determined in the beginning, there remains to be found the distance between the objective and the object, which will represent approximately the equivalent focus. The general formula is, Object, O : Image, I : : equivalent focus, F : 250. Supplying the known values, O $\frac{2}{10}$, I $\frac{10}{10}$ then $\frac{2}{10}$ m : 1 mm : : F : 250 whence F 50 mm. That is, the equivalent focus is approximately 50 millimeters.

§ 358. **Determination of Initial or Independent Magnification of the Objective.** The initial magnification means simply the magnification of the real image ($A^1 B^1$, Fig. 21) unaffected by the ocular. It may be determined experimentally exactly as described in § 357. For example, the image of the object ($\frac{1}{50}$ mm.) measured by the ocular micrometer, at a distance of 250 mm. is $|\frac{1}{10}|$ mm., *i. e.*, it is five times magnified, hence the initial magnification of the 50 mm. objective is approximately five.

Knowing the equivalent focus of an objective, one can determine its initial magnification by dividing 250 mm. by the equivalent focus in millimeters. Thus the initial magnification of a 5 mm. objective is $\frac{250}{5}$", 50 ; of a 3 mm., $\frac{250}{3}$ " 83.3 ; of a 2 mm., $\frac{250}{2}$" 125, etc.

§ 359 **Determining the Equivalent Focus of an Ocular.**—If one knows the initial magnification of the objective (§ 358) the approximate equivalent focus of the ocular can be determined as follows :

The field lens must not be removed in this case. The distance between the position of the real image, a position indicated in the ocular by a diaphragm, and the back lens of the objective should be made 250 mm., as described in § 357, 358, then by the aid of Wollaston's camera lucida the magnification of the whole microscope is obtained, as described in § 149. As the initial power of the objective is known, the power of the whole microscope must be due to that initial power multiplied by the power of the ocular, the ocular acting like a simple microscope to magnify the real image (Fig. 21).

Suppose one has a 50 mm. objective, its initial power will be approximately 5. If with this objective and an ocular of unknown equivalent focus the magnification of the whole microscope is 50, then the real image or initial power of the objective must have been multiplied 10 fold. Now if the ocular multiplies the real image 10 fold it has the same multiplying power as a simple lens of 25 mm. focus for, using the same formula as before : o 5 : i 50 : : f : 250 whence F 25. The matter as stated above is really very much more complex than this, but this gives an approximation.

For a discussion of the equivalent focus of compound lens-systems, see modern works on physics; see also C. R. Cross, on the Focal Length of Microscopic Objectives. Franklin Inst. Jour., 1870, pp. 401-402 ; Monthly Micr. Jour., 1870, pp. 149-159. J. J. Woodward, on the Nomenclature of Achromatic Objectives, Amer. Jour. Science, 1872, pp. 406-414 ; Monthly Micr. Jour., 1872, pp. 66-74. W. S. Franklin, method for determining focal lengths of microscope lenses. Physical Review, Vol. I. 1893, p. 142. See pp. 1037 to 1049 of Carpenter-Dallinger for mathematical formulæ ; also Daniell, Physics for medical students ; Czapski, Theorie der optischen Instrumente; Dippell, Nägeli und Schwendener, Zimmermann.

PREPARATION OF DIAGRAMS.

§ 360. For class room work many diagrams are needed. Those which are purchased soon get out of date, but from their cost one does not feel like throwing them away. It is a fact, however, that so much of one's education comes through the eye that it is not safe to have an incorrect diagram for students to study The visual impression is liable to outlive the verbal correction. To avoid incorrect diagrams or those that no longer represent the present state of knowledge one may use blackboard diagrams. By means of the flexible, or roll blackboards, one may make the diagrams anywhere and hang them in the lecture-room. This also is of advantage where several must use the same lecture-room.

For permanent diagrams which shall be as easy to make as the ordinary blackboard drawings and so cheap that there is no temptation to preserve them after their usefulness is past, one may adopt the suggestion of Dr. Dunnington of the University of Virginia and make the diagrams on manilla paper with ordinary blackboard crayons, and then fix them so that they will not rub by hanging them where the face cannot touch and putting the fixative on the back with a brush or a sponge. The fixative may be readily prepared by mixing one liter (1000 cc.) of ordinary painter's turpentine with 100 cc. of dammar varnish. If these are well shaken together the dammar will be dissolved in the turpentine, and then if the mixture is put on the *back* of the diagram it will soak through the paper and upon drying, will fix the crayon lines so that they will not rub. At the present day black blackboard crayons are manufactured and they are somewhat easier to shade with, and very much cheaper for all black work than the Condé crayons. The Condé crayons are better for lettering, however.

It requires only about 12 hours for the fixative to dry. Diagrams may be used in less time, but they should not be rolled much sooner. If one wishes to have rollers, and they are very convenient, they are easily made by using what the carpenters call " half-round." If two of these are used, the paper being put between, and then the sticks nailed together, very neat looking diagrams are produced. Of course if it is desired, water colors may be used upon these diagrams either before or after the crayon has been fixed.

In making diagrams from figures in books, if one desires to enlarge a definite number of times the drawing paper should be laid out in squares. If these are made lightly in pencil they will not show in the finished diagram unless one scans it closely.

DRAWINGS FOR PHOTO-ENGRAVING.

(WRITTEN BY MRS. GAGE).

§ 361. The inexpensive processes of reproducing drawings bring within the reach of every writer upon scientific subjects the possibility of presenting to the eye by diagrams and drawings the facts discussed in the text. Though artistic ability is necessary for perfect representation of an object, neatness and care will enable any one to make a simple illustrative drawing, from which an exact copy is obtained and a plate prepared for printing.

§ 362. A shaded drawing prepared by washes of India ink can be reproduced by the "half-tone process," which is the same as that used in the case of photographs. (See § 351, Figs. 160-161). The process usually called photo engraving is that by which all the line drawings in this book were reproduced, and is much less expensive than the "half-tone" process. For photo-engraving, only pure black and white can be used in the drawing, as shades of gray are not successfully reproduced.

§ 363. **Outfit for Drawings.**—A perfectly squared drawing board; a T-square; thumb tacks; a right-line pen; a circle pen; an assortment of smooth pointed pens, including for fine work a lithographic pen; very soft, hard, and medium pencils; fine scissors; fine forceps; a sharp pointed knife or scalpel; smooth, white bristol board; perfectly black ink. To test the ink draw extremely fine lines and look at them with a magnifying glass. Most of the water-proof and liquid India inks on the market answer well.

§ 364. **Size of Drawing.**—It is first necessary to decide upon the scale at which the drawings are to be made. It is always recommended that they be made large, in order that in the photo engraving they may be reduced in size, and thus lessen the apparent imperfections of the drawings. The amount of reduction must be determined by each individual, depending upon whether a fine and close or coarse and broad style is natural. In the former case a reduction of $\frac{1}{5}$ is sufficient, in the latter, $\frac{1}{3}$ or even $\frac{1}{2}$ is desirable. The most generally useful reduction is found to be $\frac{1}{3}$*.

If one knows the size of the page upon which the figure or plate is to be printed, it is easy to plan exactly as to the size of the drawings. For example, a finished plate is to be 10 x 16 cm., and the reduction is $\frac{1}{3}$, then the drawing should be made $\frac{1}{2}$ larger than the plate; that is, 15 x 24 cm. and $\frac{1}{3}$ reduction will produce the exact size needed. Then the enlarged page so determined can be outlined by the T-square and the drawings artistically arranged in the space. If one does not know at the outset the size of the page, the drawings may be made upon separate papers, closely trimmed, arranged on card board, care being taken that they do not overrun the limit of the page, as above determined, and pasted in position. This is an exceedingly practical method of procedure, even when the size of the page is known. Care should be taken to keep all straight lines representing perpendicular and horizontal directions upon the individual drawings, in correct relations to the corresponding outlines of the completed plate. For instance, the scale of magnification (§ 177) should be parallel with the bottom of the page, and this is easily determined by the T-square.

§ 365. For mechanical drawings, the exact plan is carefully plotted, dots are placed for intersections and endings of lines; the lines are lightly put in with a sharp pointed, medium pencil, and then with a right-line pen and a circle pen.

§ 366. For a large natural history object, a blue print (§ 349) or a tracing made from a camera (§ 350) is obtained; or for a microscopic object, a camera lucida or embryograph drawing is made (§ 176, 178). The outlines are carefully corrected by comparison with the object, and sharply defined. The back of this photograph or drawing is blackened with a soft lead pencil, and rubbed gently with a bit of paper to make an even coating. To transfer the drawing it is placed over the drawing paper and secured in the desired position with thumb tacks. All the outlines are then traced with an ivory point, or a very hard, round pointed lead pencil, and the outlines will appear on the drawing paper. These are again lightly retouched with a pencil to complete defective lines. If it is desired to reverse a drawing, as in the case of a tracing made with the camera (§ 350), it is placed against a well lighted window pane, its outlines followed upon the back with a pencil, and then its original outlines retraced with a soft pencil. It is placed face down upon the drawing paper and the lines upon the back traced with a hard point, and the outlines will be transferred to the drawing paper.

A. If a diagram or outline drawing is desired, the outlines thus made are followed by the pen, heavy, light, and interrupted lines being used to indicate different classes of facts, according to the special need.

B. If a shaded drawing is to be made, only those outlines are traced in ink which indicate cut edges of tissue, or for some reason need to be emphasized in an especial manner. Then the shading is put in, the deep shadows and high lights being strongly marked and the gradations carefully determined. This is more easily

---

* In any case, the amount of reduction should be clearly indicated to the photo-engraver.

done if a pencil or wash drawing is first made. The shading may either be done by dotting the surface with a pen (stipple) or by lines. It is much easier to produce acceptable results by stippling, because a slight erasure may be made or a deeper shadow obtained by adding a few carefully placed dots. This method is well adapted for showing the structure of histological specimens.

C. If lines are to be used for shading, the beginner will find it more satisfactory to use the fine lithographic pen, the line extending from the shadow toward the high light and ending in a series of dots. The deeper shadow is reinforced by a second or even third series of lines, and these should rarely cross each other at right angles. They should rather follow the contour of the surface represented and in crossing should form diamond shaped spaces. A study of good copper or steel engravings will aid one in securing a proper method and instances of excellent pen work abound in the first-class illustrated magazines from which valuable suggestions can be obtained

D. It is sometimes desirable to be able to put white lines over a dark shadow. In this case a white ink may be used.

E. Occasionally a very dark picture is needed. In this case a specially prepared paper may be used and covered with an even wash of black ink. Lines are cut out with a sharp instrument, thus leaving white lines on a black background.

F. Paper with a raised stipple is sometimes used. Upon this the shading is done by wax crayons, the crayon adhering to the elevations and leaving the depressions white. Lines and deep shadows are put in with ink. In this way more excellent drawings can be made than by the more laborious method with pen and ink, but as a rule the results of the photo engraving are not so satisfactory.

G. In case it is necessary to show different colors, the drawing is made in different colored inks exactly as desired, pale colors being avoided. The photo engraver makes a plate for each color, and the accuracy necessary in their production and printing makes the process many times more expensive than the simple reproduction of a black and white drawing.

§ 367. *Lettering.*—A half-tone engraving from a photograph may have letters placed upon it by a second printing.

The most carefully prepared drawing may be artistically ruined by placing upon it ill-formed and irregular lettering, and as even, artistic lettering is almost a trade by itself, it is recommended that either the drawing be sent to the photo-engraver with directions to have the letters properly put in, or what is more satisfactory, to have the letters, abbreviations and words needed, printed and separately placed on the drawings.

To do this, type of the proper size for the reduction determined upon is chosen. Numbers of the following size have been found convenient for numbering the separate figures of a plate for ⅓ reduction ;

60, 61, 62,

of the following size for serial parts of the individual drawings,

1, 2, 3, 4, 5,

while italic letters of the following size show what has proved most satisfactory for ⅓ reduction,

R. bc L. g. t. Filament of Necturus C. Leucocytes.

For ⅛ reduction the following are large enough,

Meten.   mlc.   mlp.   mlpr.   1 2 3 4 5.

It is convenient to have the complete alphabet both in capitals and small letters and series of numbers in different sizes. The printing should be done on firm, but light weight, white paper (20 lb. Demy has been found satisfactory).

The slips thus printed are pinned upon a board and coated upon the back with liquid gelatin (§ 311) and dried. With fine scissors the words are cut out neatly, moistened and with fine forceps placed upon the drawing and pressed down with blotting paper. Upon a large and complicated plate, the letters and words should be placed so as not to offend the eye. Those which are intended to be parallel to each other and to an edge of the plate should be strictly so, as determined in adjusting them by means of the T-square. The numbers of the different figures of a plate can be arranged so as to lend to the general harmony and still be kept close to the designated figure. If a great deal of lettering and numbering is applied a careful plan of the exact place of each word must be determined before any are fastened, otherwise confusion results. A word may be made to follow the arc of a circle, by snipping the slip on which it is printed with the scissors and curving it after it is moistened.

§ 368. After the letters and words are thus fastened, if any of them are not exactly upon the parts to which they refer, a line from the part to the word should be drawn with the right line pen and T-square. Dotted or full lines may be used, according to which will show most clearly—and in passing over a deep shadow, it may be scratched out with a sharp scalpel, leaving a white line.

§ 369. After and only after every part of the drawing is as complete as possible should the eraser be lightly used to remove pencil marks and the sharply pointed knife or scalpel to remove slight defects, as the overrunning of an inked line or to pick out shadows which are too deep. This is to avoid roughening the surface of the drawing paper and thus making it impossible to add clear cut lines with the pen.

§ 370. Considerable modification of a photo-engraving can be made by a skillful engraver with his tools. Most of the illustrations of the current magazines are half tone or photo-engravings retouched by the engraver. Slight defects of a plate, a line a trifle too long, a complete line which should be interrupted, a superfluous line can be easily remedied, but a line cannot easily be added. A deep shadow can be lightened or removed, but it is cheaper to renew the drawing than to undertake extensive changes in the plate. In the case of a plate made up of drawings on separate papers, the edge of each paper casts a shadow which would appear in the plate except that it is cut out by the photo-engraver, and in the same way the edges of the slips used for lettering cast shadows which the engraver removes. If any such lines should appear in the proof they can be indicated and removed before used in printing.

BIBLIOGRAPHY.

The books and periodicals named below in alphabetical order pertain wholly or in part to the microscope or microscopical methods. They are referred to in the text by recognizable abbreviations.

For current microscopical and histological literature, the Journal of the Royal Microscopical Society, the Index Medicus, the Zoologischer Anzeiger, and the Zeitschrift für wissenschaftliche Mikroskopie, Anatomischer Anzeiger, Biologisches Centralblatt and Physiologisches Centralblatt, and the smaller microscopical journals taken together furnish nearly a complete record.

References to books and papers published in the past may be found in the periodicals just named, in the Index Catalog of the Surgeon General's library; in the *Royal Society's Catalog of Scientific Papers*, and in the bibliographical references given in special papers. A full list of periodicals may also be found in Vol. XVI of the Index Catalog.

BOOKS.

Adams, G.—Micrographia illustrata, or the microscope explained, etc. Illustrated. 4th edition, London, 1771. Also Essays, 1787.

Ångström.—Recherches sur le spectre solaire, spectre normal du soleil. Upsala, 1868.

Anthony, Wm. A., and Brackett, C. F.—Elementary text-book of physics. 7th ed. Pp. 527, 165 Fig. New York, 1891.

Barker.—Physics. Advanced course. Pp. 902, 380 Fig. New York, 1892.

Bausch, E.—Manipulation of the microscope. Pp. 95, illustrated. Rochester, 1891.

Beale, L. S.—How to work with the microscope. 5th ed. Pp. 518, illustrated. London, 1880. Structure and methods.

Beauregard, H., et Galippe, V.—Guide de l'élève et du praticien pour les travaux pratiques de micographie, comprenant la technique et les applications du microscope à l'histologie végétale, à la physiologie, à la clinique, à la hygiène et à la médicine légale. Pp. 904, 570 Fig. Paris, 1880.

Behrens, J. W.—The microscope in botany. A guide for the microscopical investigation of vegetable substances. Translated and edited by Hervey and Ward. Pp. 466, illustrated. Boston, 1885.

Behrens, W., Kossel, A., und Schiefferdecker, P.—Das Mikroskop und die Methoden der mikroskopischen Untersuchung. Pp. 315, 193 Fig. Braunschweig, 1889 ?.

Brewster, Sir David.—A treatise on the microscope. From the 7th ed. of the Encyc. Brit., with additions. Illustrated, 1837.

Brewster, Sir David.—A treatise on optics. Illustrated. New edition. London, 1853.

Browning, J.—How to work with the micro-spectroscope. London, 1894.

Bousfield, E. C.—Guide to photo-micrography. 2d ed. Illustrated. London, 1892.

Carnoy, J. B., Le Chanoine.—La Biologie Cellulaire; Etude comparée de la cellule dans les deux règnes. Illustrated (incomplete). Paris, 1884. Structure and methods.

Carpenter, W. B.—The microscope and its revelations. 6th ed. Pp. 882, illustrated. London, and Philadelphia, 1881. Methods and structure.

Carpenter-Dallinger.—The microscope and its revelations, by the late William B. Carpenter. 7th edition, in which the first seven chapters have been entirely re-written, and the text throughout reconstructed, enlarged and revised by the Rev. W. H. Dallinger. London and Philadelphia, 1891.

This work deals very satisfactorily with the higher problems relating to the microscope and is invaluable as a work of reference.

Clark, C. H.—Practical methods in microscopy. Illustrated. Boston, 1894.

Cooke, M. C.—One thousand objects for the microscope. Pp. 1,3 London no date. See figures and brief descriptions of pretty objects for the microscope.

Crookshank, E. M. - Photography of bacteria. London and New York, 1887.

Cross and Cole.—Modern microscopy for beginners. Part I The microscope and instructions for its use. Part II. Microscopic objects, how prepared and mounted. Illustrated. 2d edition London, 1895.

Czapski, Dr. Siegfried.—Theorie der optischen Instrumente nach Abbe Illustrated Breslau, 1893.

Dana, J. D.—A system of mineralogy. Illustrated. 6th ed. New York, 1892

Daniell, A.—A text-book of the principles of physics. Illustrated. 3d ed. London, 1895.

Daniell, A.—Physics for students of medicine. Illustrated London and New York, 1896

Davis, G. E.—Practical microscopy. 3d ed. Illustrated. London, 1895.

Dippel, L. - Handbuch der allgemeinen Mikroskopie. Illustrated. 2d ed. Braunschweig, 1898.

Dippel, L.—Grundzüge der allgemeinen Mikroskopie. Pp. 521, 245 Fig. Braunschweig, 1885 Excellent discussion of the microscope and accessories.

Dodge, Charles Wright.—Introduction to elementary practical biology : a laboratory guide for high school and college students. New York, 1894.

Ebner, V. v.—Untersuchungen über die Ursachen der Anisotropie organischer Substanzen Leipzig, 1882. Large number of references.

Ellenberger, W.—Handbuch der vergleichenden Histologie und Physiologie der Haussäugethiere. Berlin, 1884 +.

Fol, H.—Lehrbuch der vergleichenden mikroskopischen Anatomie, mit Einschluss der vergleichenden Histologie und Histogenie. Illustrated (incomplete). Leipzig, 1884. Methods and structure.

Foster, Frank P.—An illustrated encyclopædic medical dictionary, being a dictionary of the technical terms used by writers on medicine and the collateral sciences in the Latin, English French and German languages. Illustrated, four quarto volumes. 1888-1893.

Fraenkel und Pfeiffer.—Atlas der Bacterien-Kunde. Berlin, 1889

Francotte, P.—Manuel de technique microscopique. Pp. 433, 110 Fig. Brussels, 1886.

Francotte, P.—Microphotographie appliquée à l'histologie, l'anatomie comparée et l'embryologie. Brussels, 1886.

Frey, H.—The microscope and microscopical technology. Translated and edited by G. R. Cutter. Pp. 624, illustrated. New York, 1880. Methods and structure.

Gage, A. P.—The principles of physics. Illustrated. Boston and London, 1895.

Gamgee, A.—A text-book of the physiological chemistry of the animal body Part I. pp. 48-63 Fig. London and New York, 1880. Part II, 1893.

Gibbs, H.—Practical histology and pathology. Pp. 107. London, 1880. Methods.

Gillay, Dr. E.—Sieben Objecte unter dem Mikroskop. Einführung in die Grundlehren der Mikroskopie. Leiden, 1893. This is also published in the Holland (Dutch) and French language.

Goodale, G. L. Physiological botany. Pp 499 36, illustrated. New York, 1885. Structure and methods.

Griffith and Henfrey.—The Micrographic Dictionary ; a guide to the examination and investigation of the structure and nature of microscopic objects. Fourth edition, by Griffith, assisted by Berkeley and Jones. London, 1883.

Gould, G. M.—The illustrated dictionary of medicine, biology and allied sciences Illustrated 3d ed., Philadelphia, 1896. This is recognized as the best single volume medical dictionary It is especially satisfactory for the worker with the microscope.

Halliburton, W. D.—A text-book of chemical physiology and pathology Pp 874, 104 illus London and New York, 1891.

Harker, A.—Petrology for students ; an introduction to the study of rocks under the microscope. Cambridge, 1895.

Harting, P.—Theorie und algemeine Beschreibung des Mikroskopes. 2nd ed. 3 vols. Braunschweig, 1866.

Hogg, J.—The microscope, its history, construction and application. New edition, illustrated. Pp. 764. London and New York, 1883. Much attention paid to the polariscope.

Huber, G. Carl—Directions for work in the histological laboratory. More especially arranged for the use of classes in the University of Michigan. 2d ed. Illustrated. Ann Arbor, 1895.

James, F. L.—Elementary microscopical technology. Part I, the technical history of a slide from the crude material to the finished mount. Pp. 107, illustrated. St. Louis, 1887.

Jeserich, P.—Die Mikrophotographie auf Bromsilbergelatine bei natürlichem und künstlichem Lichte. Figs. and Plates. Pp. 245. Berlin, 1888.

King, J.—The microscopist's companion. A popular manual of practical microscopy. Illustrated. Cincinnati, 1859.

Klément and Renard.—Réactions microchemiques à cristaux et leur application en analyse qualitative. Pp. 126, 8 plates. Bruxelles, 1886.

Kraus, G.—Zur Kentniss der Chlorophyllfarbstoffe. Stuttgart, 1872.

Latteux, P.—Manuel de technique microscopique. 3d ed, Paris, 1887.

Le Conte, Joseph.—Sight—an exposition of the principles of monocular and binocular vision. Pp. 275, illustrated. New York, 1881.

Lee, A. B.—The microtomist's vade-mecum. A hand-book of the methods of microscopic anatomy. 4th ed. Philadelphia, 1896.

Lehmann, C. G.—Physiological chemistry. 2 vols. Pp. 648 : 547, illustrated. Philadelphia, 1855.

Lehmann, O.—Molekularphysik mit besonderer Berücksichtigung mikroskopischer Untersuchungen und Anleitung zu solchen, sowie einem Anhang über mikroskopische Analyse. 2 vols. Illustrated. Leipzig, 1888-1889.

Lockyer, J. N.—The spectroscope and its application. Pp. 117, illustrated. London and New York, 1873.

M'Kendrick, J. G.—A text-book of physiology. Vol. I, general physiology. Pp. 516, 318 illus. New York, 1888.

Macdonald, J. D.—A guide to the microscopical examination of drinking water. Illustrated. London, 1875. Methods and descriptions.

MacMunn, C. A.—The spectroscope in medicine. Pp. 325, illustrated. London, 1885.

Martin, John H.—A manual of microscopic mounting with notes on the collection and examination of objects. 2d ed. Illustrated. London, 1878.

Mason, John J.—Minute structure of the central nervous system of certain reptiles and batrachians of America. Illustrated by permanent photo-micrographs. Newport, 1879-82.

Matthews, C. G., and Lott, F. E. The microscope in the brewery and the malthouse. Illustrated. London, 1889.

Mayall, Jr., John.—Cantor lectures on the microscope, delivered before the society for the encouragement of arts, manufactures and commerce. Nov.-Dec., 1885. History of the microscope, and figures of many of the forms used at various times.

Moitessier, A.—La photographie appliquée aux recherches micrographiques. Paris, 1866.

Nägeli und Schwendener.—Das Mikroskop, Theorie und Anwendung desselben 2d ed. Pp. 647, illustrated. Leipzig, 1877.

Neuhaus, R.—Lehrbuch der Mikro-photographie. Illustrated. Braunschweig, 1890.

Pelletan, J.—Le microscope, son emploi et ses applications. 278 figures, 4 plates. Paris, 1878.

Phin, J.—Practical hints on the selection and use of the microscope for beginners. 6th ed. Illustrated. New York, 1891.

Preyer, W.—Die Blutkrystalle. Jena, 1871. Full bibliography to that date.

Pringle, A.—Practical photo-micrography. Pp. 193, illustrated. New York, 1890.

Procter, R. A.—The spectroscope and its work. London, 1882.

Queckett, J.—A practical treatise on the use of the microscope, including the different methods of preparing and examining animal, vegetable and mineral structures. Pp. 515, 12 plates. 2d ed. London, 1852.

Ranvier, L.—Traité technique d'histologie. Pp. 1109, illustrated. Paris, 1 7 -1. Structure and methods. Also German translation 1888.

Reeves, J. E.—A hand-book of medical microscopy, including chapters on bacteriology, neoplasms and urinary examination. Illustrated. Philadelphia, 1894.

Reference Hand-Book of the medical sciences. Albert H. Buck, editor. 8 quarto vols. Illustrated with many plates, and figures in the text. New York, 1885-1889. Supplement 1893.

Richardson, J. G.—A hand-book of medical microscopy. Pp. 333, illustrated. Philadelphia, 1871. Methods and descriptions.

Robin, Ch.—Traité du microscope et des injections. 2d ed. Pp. 1101, illustrated. Paris, 1877. Methods and structure.

Roscoe, Sir Henry.—Lectures on spectrum analysis, 4th ed. London, 1885.

Rosenbusch, K. H. F., translated by Iddings, P.—Microscopical physiography of the rock making minerals. Illustrated. New York, 1889.

Ross, Andrew.—The microscope. Being the article contributed to the "Penny Cyclopaedia." Republished in New York in 1877. Illustrated.

Rutherford, W.—Outlines of practical histology. 2d ed. Illustrated. Pp. 194. London and Philadelphia, 1876. Methods and structure.

Satterthwaite, F. E. (editor).—A manual of histology. Pp. 478, illustrated. New York, 1881. Structure and methods.

Schäfer, E. A.—A course of practical histology, being an introduction to the use of the microscope. Pp. 304, 40 Fig. Philadelphia, 1877. Methods.

Schacht, H., translated by F. Currey.—The microscope in its special application to vegetable anatomy and physiology. Illustrated. London, 1853.

Schellen, H.—Spectrum analysis, translated by Jane and Caroline Lassell. Edited with notes by W. Huggins. 13 plates, including Angström's and Kirchhoff's maps. London, 1885.

Science Lectures at South Kensington. 2 vols. Pp. 290 and 314, illustrated. One lecture on microscopes and one on polarized light. London, 1878-1879.

Sedgwick and Wilson.—General biology. 2d ed. Illustrated. New York, 1895.

Seiler, C.—Compendium of microscopical technology. A guide to physicians and students in the preparation of histological and pathological specimens. Pp. 130, illustrated. New York, 1881.

Silliman, Benj., Jr.—Principles of physics, or natural philosophy. 2d ed., rewritten. Pp. 710, 722 illustrations. New York and Chicago, 1860.

Starr, Allen M., with the coöperation of Oliver S. Strong and Edward Leaming.—An Atlas of nerve cells. Columbia University Press, New York, 1896. The atlas consists of text, diagrams, and some of the best photo-micrographs that have ever been published.

Sternberg, G. M.—Photo-micrographs, and how to make them. Pp. 204, 20 plates. Boston, 1883.

Sternberg, G. M.—A manual of bacteriology. Pp. 886, 268 Fig. viii plates. New York, 1892. Bibliography on photographing bacteria.

Stokes, A.—Microscopical praxis. Illustrated. Portland, Conn., 1894.

Stokes, A.—Aquatic microscopy for beginners, or common objects from the ponds and ditches. Illustrated. Portland, Conn., 1896.

Stowell, Chas. H.—The students' manual of histology, for the use of students, practitioners and microscopists. 3d ed. Pp. 368. Illustrated. Ann Arbor, 1884. Structure and methods.

Strasburger, E.—Das botanische Practicum. Anleitung zum Selbststudium der mikroskopischen Botanik, für Anfänger und Fortgeschrittnere. Pp. 664, illustrated. Structure and methods. Also English translation.

Suffolk, W. T.—On microscopical manipulation. 2d ed. Pp. 227, illustrated. London 1870.

Suffolk, W. T.—Spectrum analysis applied to the microscope. Referred to in Beale.

Thomas, Mason H., and Wm. R. Dudley.—A laboratory manual of plant histology. Illustrated. Crawfordsville, Ind., 1894.

Trelease, Wm.—Poulsen's botanical micro-chemistry; an introduction to the study of vegetable histology. Pp. 118. Boston, 1884. Methods.

Trutat, M.—La photographie appliquée à l'histoire naturelle. Pp. 228, illustrated. Paris, 1884.

Valentin, G.—Die Untersuchung der Pflanzen und der Thiergewebe in polarisirtem Licht. Leipzig, 1861.

Vierord: Die quantitative Spectralanalyse in ihrer Anwendung auf Physiologie. 1876.
Vogel, Conrad.—Practical pocket-book of photography. Pp. 202, Figs. London, 1893.
Vogel, H. W.—Practische Spectralanalyse irdischer stoffe ; Anleitung zur Benutzung der Spectralapparate in der qualitativen und quantitativen chemische Analyse organischer und unorganischer Körper. 2d ed. Figs. Berlin, 1889.
Wethered, M. -Medical microscopy. Pp. 406, Figs. London and Philadelphia, 1892.
Whitman, C. O.—Methods of research in microscopical anatomy and embryology. Pp. 255, illustrated. Boston, 1885.
Wilder and Gage.—Anatomical technology as applied to the domestic cat. An introduction to human, veterinary and comparative anatomy. Pp. 575, 130 Fig. 2d ed. New York and Chicago, 1886.
Wilson, Edmund B., with the coöperation of Edward Leaming.—An atlas of fertilization and Karyokinesis. Columbia University Press, New York, 1895. This atlas marks an era in embryological stu ly. It has admirable text and diagrams, but the distinguishing feature is the large number of almost perfect photo-micrographs.
Wood, J. G.—Common objects for the microscope. Pp. 132. London, no date. Upwards of 400 figures of pretty objects for the microscope, also brief descriptions and directions for preparation.
Wormly, T. G.—The micro-chemistry of poisons. 2d ed. Pp. 742, illus. Philadelphia, 1885.
Wythe, J. H.—The microscopist ; a manual of microscopy and a compendium of microscopical science. 4th ed. Pp. 434, 252 Fig. Philadelphia, 1880.
Zeiss, R.—Special-catalog über Apparate für Mikro-photographie. Jena. Instructions for using the apochromatic objectives and projection oculars, etc.
Zimmermann, Dr. A.—Das Mikroskop, ein Leitfaden der wissenschaftlichen Mikroskopie. Illustrated. Leipzig und Wein, 1895.
See also Watt's chemical dictionary, and the various general and technical encyclopedias.

PERIODICALS.*

The American journal of microscopy and popular science. Illustrated. New York, 1876-1881.
The American monthly microscopical journal. Illustrated, Washington, D. C., 1880 ÷.
American naturalist. Illustrated. Salem and Philadelphia, 1867 ÷.
American quarterly microscopial journal, containing the transactions of the New York microscopial society. Illustrated. New York, 1878.
American microscopical society, Proceedings. 1878 ÷.
Anatomischer Anzeiger. Centralblatt für die gesammte wissenschaftliche Anatomie. Amtliches Organ der anatomischen Gesellschaft. Herausgegeben von Dr. Karl Bardeleben. Jena, 1886 ÷. Besides articles relating to the microscope or histology, a full record of current anatomical literature is given.
Annales de la société belge de microscopie. Bruxelles, 1874 ÷.
Archiv für microscopische Anatomie. Illustrated. Bonn, 1865 ÷.
Centralblatt für Physiologie. Unter Mitwirkung der physiologischen Gesellschaft zu Berlin Herausgegeben von S. Exner und J. Gad. Leipzig und Wien, 1887 ÷. Brief extracts of papers having a physiological bearing. Full bibliography of current literature.
English mechanic. London, 1866 ÷. Contains many of the papers of Mr. Nelson on lighting, photo-micrography, etc.
Index Medicus. New York, 1879 ÷. Bibliography, including histology and microscopy.
International journal of microscopy and popular science. London, 1890 ÷.
Journal of anatomy and physiology. Illustrated. London and Cambridge, 1867 ÷.
Journal de micrographie. Illustrated. Paris, 1877-1892.
Journal of microscopy and natural science. London, 1885 +.

*NOTE.—When a periodical is no longer published, the dates of the first and last volumes are given ; but if still being published, the date of the first volume is followed by a plus sign.
See Vol. XVI of the Index Catalog of the Library of the Surgeon General's office for a full list of periodicals.

Journal of the New York microscopical society. Illustrated. New York.
Journal of physiology. Illustrated. London and Cambridge, 1878-
Journal of the American chemical society. New York, 1879-
Journal of the chemical society. London, 1848-.
Journal of the royal microscopical society. Illustrated. London, 1878-. Bibliography of works and papers relating to the microscope, microscopical methods and histology. It also includes a summary of many of the papers.
Journal of the Queckett microscopical club. London, 1868-
The Lens, a quarterly journal of microscopy, and the allied natural sciences, with the transactions of the state microscopical society of Illinois. Illustrated. Chicago, 1872-1873.
The Microscope. Illustrated. Washington, D. C., 1881-
Microscopical bulletin and science news. Illustrated. Philadelphia, 1883-. The editor Edward Pennock introduced the term "par-focal" for oculars (see vol. iii, p 31)
Monthly microscopical journal. Illustrated. London, 1869-1877.
Nature. Illustrated. London, 1869-.
The Observer. Portland, Conn., 1890-.
Philosophical Transactions of the Royal Society of London. Illustrated. London. 1665-
Proceedings of the American microscopical society, 1878-.
Proceedings of the Royal Society. London, 1854-.
Quarterly journal of microscopical science. Illustrated. London, 1853-.
Science Record. Boston, 1883-4.
Zeitschrift für Instrumentenkunde. Berlin, 1881-.
Zeitschrift für physiologische Chemie. Strassburg, 1877-.
Zeitschrift für wissenschaftliche Mikroskpie und für mikroskopische Technik. Illustrated. Braunschweig. 1884-. Methods, bibliography and original papers.

Besides the above-named periodicals, articles on the microscope or the application of the microscope appear occasionally in nearly all of the scientific journals. One is likely to get references to these articles through the Jour. Roy. Micr. Soc. or the Zeit. wiss. Mikroskopie Excellent articles on Photo-micrography occur in the special Journals and Annuals of Photography.

INDEX.

A

Abbe apertometer, 212.
Abbe camera lucida, 111; arrangement of, 112; drawing with, 115-118; figures, 98, 110, 114; hinge for prism, 117; inclined microscope with, 115.
Abbe condenser or illuminator, 44, 47; amount of light for, 43; dark-ground illumination with, 48; experiments, 46; laboratory microscope with, 61; light, axial and oblique, 46; mirror with, 46.
Abbe's test-plate, method of using, 210.
Aberration, chromatic, 4; by cover glass, 51; negative, 52; spherical, 4, 210.
Absorption spectra 122, 135; amount of material necessary and its proper manipulation, 130; Angström & Stokes' law of, 123; banded, not given by all colored objects, 133; of blood, 131; of colored bodies, 123, 133; of colorless bodies, 134; of minerals, 134; of permanganate of potash, 131.
Acetylene light, 36, 48.
Achromatic condenser, 40, 41; objectives, 12, 61; oculars, 23; triplet, 7.
Achromatism, 13.
Actinic focus, 189; image, 184.
Adjustable objectives, 12, 13, 52; experiments, 51; and micrometry, 106; and photo-micrography, 201.
Adjusting collar, graduation of, 52.
Adjustment, of analyzer, 136; coarse or rapid, 60; fine, 60; focusing, 60; of objective, 12, 13, 51; of objective for cover glass, specific directions, 53; testing, 60; with graduated collar, 53.
Aerial image, 30, 31.
Air bubbles, 84, 85; with central and oblique illumination, 84, 85; air and oil, distinguished optically, 85; by reflected light, 85.
Albumen fixative, Mayer's, 175.
Albuminous material, removal of, 58.
Alcohol, ethylic, 175.
Alcoholic dye, staining sections with, 167.
Alum solution, 175.
Amici prism, 120, 126.
Amplifier, 97, 214.
Amplification of microscope, 92.
Analyzer, 126, 136; adjustment and putting in position, 136.
Anastigmatic objective, 196.
Angle of aperture, 16, 17.
Angström and Stokes' law of absorption spectra, 123.
Angular aperture, 16, 17.
Anisotropic, 137.
Apertometer, Abbe's, 212.
Aperture of objective, 17-22, 212; angular, 16, 17; formula for, 17, 18; of illuminating cone, 44; numerical, 17, 20, 212; numerical of condenser, 43; and optical section, 87; significance of, 20.
Aplanatic cone, 44; objectives, 12; ocular, 23.
Apochromatic condenser, 40; objectives, 12, 61, 195.
Apparatus and material, 1, 33, 80, 92, 109, 120, 140, 183; for drawing, 216; for photo-micrography, 185, 191.
Appearances, interpretation, 80, 91.
Aristotype paper, 208.
Arrangement of condenser, 45; of lamp, bull's eye and microscope, 49; minute objects, 180; serial sections, 169; tissue for sections, 168.
Artifacts, 81.
Artificial illumination, 36, 46, 48; for photo-micrography, 200.
Autograph, photo-micrographic camera, 190-192.
Avoidance of diffusion currents, 162; of distortion, 111.
Axial light, 35, 84; experiments, 39; with Abbe illuminator, 46.
Axial point, 16; ray, 35.
Axis, optic, 2, 3, 11; of illuminator, 47, secondary, 5, 47.

B

Back combination or system of objective, 10-13.
Bacillus tuberculosis, 56.
Balsam, 175; bottle, 164; balsam, mounting in, 163, 167; preparation of, 175. removal from lenses, 58, natural, 176; neutral, 176; removal from slides, 141; xylene, 175, 176.
Banded absorption spectra not given by all colored objects, 133.
Bands, absorption, 123.
Base, circular, of microscope, 73.
Bed, camera, 207.

Benzin, removing of from sections, 166.
B axial crystals, 138.
Bibliography, 32, 91, 108, 119, 135, 139, 173, 182, 209, 215.
Binocular vision, for getting magnification, 93.
Blood, absorption spectrum of, 131; or other albuminous material, removal, 58; velocity of current, 88.
Body of microscope, Frontispiece.
Bottle for balsam, glycerin or shellac, 164.
Bowl, waste, 162.
Bread crumbs, examination of, 90.
Bromide paper, 208.
Brownian movement, 88.
Brunswick black, removal from lenses,58.
Bubble, air, 84-85.
Bull's-eye, 49, 191, 200; engraving, glass for, 198.
Burning point, 6; finding of, 30.
Butterfly scales, 90.

C

Cabinet for microscopical preparations, 173, 174.
Calipers, micrometer, 143, 141; pocket, 143.
Camera bed, 207; photographic, for drawing, 208; long and short bellows, 187, 200; photo-micrographic, 186, 188, 190-193, 202, 205, 207; testing, 185; vertical, 188, 293, 202, 205, 207.
Camera lucida, Abbe, 111; arrangement of, 112, 113; drawing with, 114, 115, 118; hinge for prism, 112; with inclined microscope, 115; laboratory microscope with, 61; magnification of microscope with, 92, 118.
Camera lucida, definition, 109; figures of, 98, 110 111, 114; Wollaston's, 95, 111.
Canada balsam, 175; mounting in, 163, 167; preparation of, 175; removal from lenses, 58; removal from slides, 141.
Carbol-xylene, 176.
Carbon-monoxide hemaglobin, spectrum of, 133.
Carbonate of lime, pedesis, 89.
Card catalog, 173; centering, 149.
Care of, eyes, 58; microscope, mechanical parts, 57; optical parts, 57; negatives, 197; water immersion objectives, 55.
Carmine to show currents and pedesis, 89; spectrum of, 133.
Castor-xylene clarifier, 176.
Catalog, cards, labels, ink for, 173.
Cataloging, formula, 172; preparations, 171-172.
Celloidin for coating glass rod, 87.

Cells, deep, thin, 148; isolated, preparation of, 155; mounting, 148; staining, 155.
Cement, shellac, 179.
Cementing collodion, 177.
Center, optical, 2, 3.
Centering and arrangement of illuminator, 41, 45; card, 149; condenser for photo-micrography, 201; diaphragm, 42; image of source of illumination, 42
Central light, 35, 84; with a mirror, 39.
Chain lens holder, 199
Chamber, moist, 151.
Chimney, metal for lamp, 48
Chloroform, paraffin, 176; infiltrating with, 165; saturating with, 165.
Chromatic aberration, 4; correction, 12, 13; correction, test for, 210.
Chemical focus, 13; rays, 13.
Clarifier, 176; castor-xylene. 176.
Cleaning back lens of objective, 58; homogeneous objectives, 56; mixtures for glass, 145; slides and cover-glasses, 140-142; water immersion objectives, 55.
Clearer, clearing, 153, 164, 167, 176.
Clearing mixture, preparation of, 176.
Clinical microscope, 76.
Clothes moth, examination of scales, 90.
Cloudiness of objective and ocular, how to determine, 82; removal, 58.
Coarse adjustment of microscope, Frontispiece; testing. 60.
Cob-web micrometer, 104.
Collective, 24
Collodion, 176; for coating glass rod, 87; cementing, 177; clarifying, 158; cotton, 176; hardening, 158; method, 157, 164; thick, 158, 177; thin, 157, 177.
Color correction, 13; images, 51, 56; law of, 124; production of, 138; screens, 191, 50.
Colored minerals, absorption spectra of, 134; substances, spectra of, 133.
Colorless bodies, spectra of, 134.
Coma, 53.
Combination of lenses, back and front, 10-12, 52; optical, 105.
Comparison prism, 127, 128; spectrum, 128.
Compensating ocular, 24-26.
Complementary spectra, 124.
Compound microscope, see under microscope.
Concave lenses, 3; mirror, use of, 36.
Condenser, 39-49; Abbe, 44-48; achromatic, 40, 191, 201; apochromatic, 40, 191; bull's-eye, 49, 198, 200; centering, 41, 45; illuminating cone with, 44; mirror for, 46; non-achromatic,

44; numerical aperture of, 43 44;
optic axis of, 41, 45; for photo mi-
crography, 40, 191, 201; substage, 39
40. See also illuminator.
Condensing lens, 35.
Cone, aplanatic, 44; illuminating, 43.
Conjugate foci, 3.
Construction of images, geometrical, 5.
Continuous spectrum, 123.
Contoured, doubly, 87.
Converging lens, 3, 5; lens-system, 10.
Convex lenses, 3, 5.
Corn starch, examination of, 90.
Correction, chromatic, or color, 13, 210;
cover-glass, 15, 52 53; cover, tube-
length for, 53-54; over and under,
13.
Cost of microscope, 63.
Cotton, collodion, 176; examination of,
90; gun, 176; soluble, 176.
Cover glass, or covering glass, 141; aberration by, 91; adjustment, specific directions, 53; adjustment for,
in pho o micrography, 201; adjustment and tube-length, 13, 14, 53;
anchoring, 150; cleaning, 141, 142;
correction, 52, 53; effect on rays
from object, 52; gauges, 143-145;
larger than object, 146; measurer,
143-145; measuring thickness of,
143; non adjustable objectives, table
of thickness 15; No 1, variation of
thickness, 143; putting on, 84, 146;
sealing, 148, 150; size of, 146; thickness of, 14, 15, 143, 170; tube-length
with, 14, 53, 54; wiping, 142; with
serial sections, 170.
Crayons, 216.
Crystals, biaxial, depolarizing, 138; from
frog for pedesis, 89.
Crystallization under microscope, 180.
Crystallography, 48, 180; list of substances for, 181
Currents, diffusion, avoidance of, 162;
in liquids, 88.
Cutting sections, 160, 166.

D

Dammar, for fixing crayon drawings,
216; removal from lenses, 58.
Dark ground illumination, 36, 47, 48;
with Abbe illuminator, 48; with
mirror, 48
Daylight, lighting with, 34.
Defining power, 21.
Dehydration, 157, 162, 167.
Demonstration microscope, 77.
Depolarizing crystals, 138.

Designation of oculars, 26, of wave
length, 129.
Determination of field of microscope, 9
magnification, 92, 94, 214, of sedi
ment in water, 180; of working dis
tance, 38.
Developers for photo-micrography, 197.
Development of negative, 196.
Diagrams with photographic objectives,
208; preparation of, 215.
Diamond, writing, 173.
Diaphragms and their employment, 35,
44; central stop, 36, 47; eccentric,
42, 48; iris, 40; ocular, 28; pin-hole,
41, 45,; size and position of opening,
35, 44; with condenser, 44.
Diffraction, grating, 122; illusory appearances due to, 91.
Diffusion currents, avoidance of, 162.
Direct, light, 34; vision spectroscope.
120.
Dispersing prism, 122.
Displacements, in mounting objects in
resinous media, 153.
Dissecting microscope, 9.
Dissociator, formaldehyde, 177; nitric
acid, 179.
Distance, principal focal, 3, 30; standard
at which the virtual image is measured, 97; working d. of simple mi
croscope or objective, 38; working
d. of compound microscope, 11, 33,
38.
Distinctness of outline, 85.
Distortion in drawing, avoidance of, 111.
Diverging lens, 3.
Dividers, measuring spread of, 93.
Double spectrum, 128; vision, 92, 93.
Doubly contoured, 87; refracting, 137.
Draw-tube, Frontispiece; pushing in, 37.
Drawing, with Abbe camera lucida, 114
115; board for Abbe camera lucida,
115-116; distortion avoidance of, 111,
embryograph for, 119; with microscope, 109; photographic camera
for, 119; for photo engraving, 216,
scale and enlargement of, 118; tran
ferring, 217.
Dry objectives, 11, 17-19; for laboratory
microscope, 61; light utilized, 18;
dry mounting, 147; numerical aperture, 18; dry plates, discovery by
Maddox, 184.
Drying negatives, rack for, 203.
Dust, of living rooms, examination of,
90; on objectives and oculars, how
to determine, 82; removal, 58.
Dye, general, staining with, 167, que
ons, 161, 167; alcoholic, 161, 167.

E

Eccentric diaphragm, 42, 48.
Embryograph, 119.
Engraving glass for bull's eye condenser, 198.
Erect image, 1.
Equivalent focal length, 11, 214; focus of objectives, 11, 26, 214; focus of ocular, 26, 214.
Elements, histological, isolation of, 154.
Eosin, 177.
Ether, sulphuric, 177.
Ether-alcohol, 177; saturated with, 157.
Ethylic alcohol, 175.
Examination of dust of living rooms, bread crumbs, corn starch, fibres of cotton, linen, silk, human and animal hairs, potato, rice, scales of butterflies and moths, wheat, 90.
Experiments. Abbe illuminator, 46; with adjustable and immersion objectives, 51; compound microscope, 26; crystallography and micro chemistry, 180; homogeneous immersion objective, 55; lighting and focusing, 36; in micro-chemistry, 180; with micro-spectroscope, 131; with micropolariscope, 137; in mounting, 146-161; photo-micrography, 193; simple microscope, 6.
Exposure of photographic plates, 194, 199, 200, 203.
Extraordinary ray of polarized light, 136
Eye and microscope, 1, 6, 10, 31.
Eyes, care of, 58; musce volitantes of, 90
Eye-lens of the ocular, 22.
Eye-piece, 22; micrometer, 102.
Eye-point, 7, 22, 110; of ocular, demonstration, 32.
Eye-shade, Ward's, 59; double, 59.

F

Farrant's solution, in mounting objects, 151; preparation of, 177.
Feather, examination of, 90.
Fibers, examination of, 90
Field, 27; with camera lucida, 95; illumination of, 44, 49; with orthoscopic ocular, 23; with periscopic ocular, 24; of view with microscope, 27, 29, 93, 109; size of, with different objectives and oculars, 27, 29
Field-lens, of ocular, 22; action of, 31; dust on, 82.
Filar micrometer ocular, 23; ocular micrometer, 103.
Filter paper, Japanese, 57.
Filters, light, 191.
Filtering balsam, etc., paper funnel for, 175.

Fine adjustment, Frontispiece; testing, 60
Fir, balsam of, 175.
Fixative, albumen, Mayer's, 175.
Fixing, reagents for, 175; tissue, 157.
Flame, image of, 43
Fluid, immersion, 55, 56; testing, 213.
Focal distance, or point, principal, 30; length, equivalent, 11.
Focus, 6; always up, 38; actinic, 189; chemical, 13; conjugate, 3; of objectives, equivalent, 11, 26, 214; of oculars, equivalent, 26, 214; optical, 13; principal, 3, 5; visual, 189
Focusing, 6, 33; adjustments, testing, 60; with compound microscope, 33; experiments, 36; glass, 197-198; with high objectives, 37; with low objectives, 36; objective for micro-spectroscope, 130; for photo micrography, 191, 199, 201; screen for photo micrography, 194-195; with simple microscope, 31; slit of micro spectroscope, 131.
Form of objects, determination of, 83.
Formal, 177.
Formaldehyde dissociator, 177; for isolation, 154.
Formula for cataloging, 172.
Fraunhofer lines, 123.
Front combination or lens of objective, 10, 11.
Frontal sections, 170.
Function of objective, 29 30; of ocular, 30.
Funnel, paper, 175.

G

Gauge, cover-glass, 143, 145.
Gelatin, liquid, preparation of, 178.
Geometrical construction of images, 5.
Glass, cleaning mixture for, 145; focusing, 197-198; ground, 29; for objectives, 12, 63; rod appearance under microscope, 86, 87; slides or slips, 140.
Glue, liquid, preparation of, 178.
Glycerin, bottle for, 164; mounting objects in, order of procedure, 149, 177; removal, 58.
Glycerin jelly, mounting objects in, order of procedure, 150; preparation of, 177
Gold size, removal from lenses, 58.
Goniometer ocular, 23.
Graduation of adjusting collar, 53.
Grating, diffraction, 122.
Ground glass, preparation of, 29.
Gun cotton, 176.

H

Hairs, examination of, 90.
Half-tones from photo-micrographs, 208, 216.
Hardening collodion, 158; tissue, 157, 164.
Heliostat, 209.
Hematoxylin, 178.
Hemoglobin spectrum, 132, 133.
Herapath's method of determining minute quantities of quinine, 181.
Histological elements, isolation of, 154.
Histology, physiological, 173.
History of photo-micrography, 184.
Holder, lens, 8; needle, 146.
Homogeneous immersion objective, 17 19; cleaning, 56; experiments, 55; for laboratory microscope, 61; light utilized, 18, 21; numerical aperture, 21.
Homogeneous liquid, 12, 213; tester for, 55, 213.
Huygenian ocular, 22, 24, 31.

I—J

Illuminating, cone, aperture of, 43, 44; for condenser, 44; objective, 14; power, 21, 22.
Illumination, for Abbe camera lucida, 116; artificial, 36, 48; artificial for photo-micrography, 194; centering image of source of, 41-45; central with air and oil, 84, 85; dark-ground, 36, 47, 48; daylight, 34; of entire field, 49; lamp for, 48; methods of, 34, 47; for micro-polariscope, 137; for micro-spectroscope, 129; oblique, with air and oil, 84, 85; for photo-micrography, 192; for Wollaston's camera lucida, 111.
Illuminator, 39-40; Abbe, 44, 47; Abbe, axial and oblique light, 36; Abbe, experiments, 46; Abbe, mirror and light for, 46; achromatic, 40; centering and arrangement, 45; immersion, 45. See also condenser.
Image, actinic, 184; aerial, 30, 31; centering i. of source of illumination, 42; color, 49, 51, 56; with dry apochromatic and water immersion objectives, 13; erect, 1; inverted, 7; inverted, real of objective, 20; of flame, 43; formation of, 3; geometrical construction of, 5; and object, size and position, 5, 11, 96; real, 5, 10, 22, 29 31, 92; refraction, 49, 55; retinal, 6, 10, 31; swaying of, 16; virtual i. and standard distance at which measured, 6, 31, 92.
Image-power of objectives, 19.

Imbedding, 158, 165.
Immersion fluid, 55, 213, illuminator, 45; liquid, 55; objective, 12, 55, 56, 61.
Incandescence or line spectra, 123.
Incident light, 34.
Index of refraction, 50; of medium in front of objective, 17-18.
Infiltration with chloroform, 165; collodion, 157, 158; ether, 157, paraffin, 165.
Initial magnification, 214.
Ink for labels and catalogs, 173; for drawing, 216.
Interpretation of appearances under the microscope, 80-91.
Inverted image, 7, 29.
Iris diaphragm, 40, 181.
Irrigating with reagents, 150.
Isochromatic plates, 194, 197.
Isolation, 154; with formaldehyde, 154 nitric acid, 155.
Isotropic, 137.
Japanese filter or tissue paper, 57, 160.
Jelly, glycerin, 177.
Jena glass, 63.
Jurisprudence, micrometry in, 108.

L

Labels and catalogs, ink for 173; preparation of, 179.
Labeling microscopical preparations, 171; photographic negative, 197 serial sections 171.
Laboratory compound microscope, 61.
Lamp, microscope, 48; spirit, 164.
Lamp-light, 36.
Law of color, 124.
Lens, concave, 3; converging, 3; convex, 3; eye, 22; field, 22, 26; holder, 8, 155, 156, 199; paper, 57; system, 10.
Lens-systems, 10.
Letters in stairs, 82; for photo-engraving, 218.
Lettering oculars, 26.
Light, with Abbe illuminator, 46; acetylene, 36, 48; artificial, 46; axial, 35, 39; axial with Abbe illuminator, 46; direct, 34; central, 35, 39; filters, 191; incident, 34; with mirror, 39 oblique, 35, 39; oblique with Abbe illuminator, 46; lamp, 36 for photo-micrography, 192; polarized, 136 reflected incident or direct, 34; transmitted, 35; utilized with different objectives, 13; wave length of, 124.
Lighting, 34; for Abbe camera lucida 115, artificial, 48; experiments, 34 and focusing, 35, 36, for micro polariscope, 137; for micro-spectro

scope, 129; with a mirror, 36; with daylight, 34; for photo-micrography, 192.
Line spectrum, 123.
Linen, examination of, 90.
Liquid, currents in, 88; homogeneous, 55, 213.

M

Macro-photography, 184.
Magnification of compensating oculars, 26; effect of adjusting objective, 105; determination of, 92, 94; expressed in diameters, 92; initial or independent, 214; method of binocular or double vision in obtaining, 93; of microscope, 92; of microscope with Abbe camera lucida, 118; of microscope, compound, 94; of microscope, simple, 93; of photo-micrographs, determination of, 202; real images, 92; table of, with ocular micrometer, 99; varying with compound microscope, 97; and velocity, 88.
Magnifier, tripod, 7, 198.
Marker for preparations, 63, 64.
Marking objects, 63, 64, 94, 101; negatives, 197; objectives, 65.
Material and apparatus, 1, 33, 80, 92, 120, 140, 175-181, 183, 210, 216.
Measure, unit of, in micrometry, 100; of wave length, 128.
Measurer, cover-glass, 143-145.
Measuring the spread of dividers, 93; thickness of cover-glass, 143.
Mechanical parts of compound microscope, 61; of microscope, care of, 57; testing, 60
Mechanical stage, 61, 65-66.
Medium, mounting, 147.
Met-hemaglobin, spectrum of, 122, 133.
Methods, collodion, 157; paraffin, 164.
Micro-chemistry, 180.
Micrometer, 92; calipers, 143, 144; cobweb, 104; filar m. ocular, 103; filling lines of, 94; net, 113; lines, arrangement of ocular and stage, 107; object or objective, 94; ocular or eye-piece, 102, 103; ocular, micrometry with, 105; ocular, ratio, 106; ocular, valuation of, 103; ocular, varying valuation of, 105; screw ocular, 103; stage, 94; table of magnification, 99.
Micrometry, definition, 100-102; with adjustable objectives, 106; comparison of methods, 107; with compound microscope, 100; and jurisprudence, 108; limit of accuracy in, 107; with ocular micrometer, 105; with simple

microscope, 100; remarks on, 106; unit of measure in, 100.
Micro-millimeter, 101.
Micron, 100; for measuring wave-length of light, 129.
Micro-photograph, 183.
Micro-photography, distinguished from photo-micrography, 183.
Micro-polariscope, 89, 136; experiments, 137; for laboratory microscope, lighting for, 137; objectives to use with, 136; purpose of, 137; selenite plate with, 138; sulphonal with, 139.
Micro-polarizer, 136.
Microscope, definition, 1; amplification of, 92; clinical, 76; demonstration, 77; dissecting, 9; care of, 57; eye and, 1, 6, 10; field of, 27, 29; focusing, 31; magnification, 92; for photomicrography, 187; polarizing, 136; price of, 61, 63; putting an object under, 27; screen, 56.
Microscope compound, definition, 7; drawing with, 110; figures, 10, 65; focusing, 36-37; for laboratory, 61; lamp, 48; magnification or magnifying power, 94; magnification and size of drawing with Abbe camera lucida, 118; mechanical parts of, 61; micrometry with, 100; optic axis of, 10; optical parts of, 10, 61; polarizing, pedesis with, 89; varying magnification, 97; working distance of, 38; testing, 60.
Microscope, simple, definition, 1; experiments with, 6; figures, 6-8, 31, 155-156, 197-199; focusing with, 33; magnification of, 93; micrometry with, 100; working distance of, 33.
Microscopic objective, 10; objective low, attached to camera, 198; objects, drawing, 109; ocular, 22; slides or slips, 140.
Microscopical preparations, cabinet for, 174; cataloging, 172; labeling, 171; mounting, 146-170; tube-length, 14, 15.
Microscopy and photography, 183.
Micro-spectroscope, 120-121; adjusting, 125; experiments, 131; focusing, 130; focusing the slit, 125; for laboratory microscope, 61; lighting for, 129; objectives to use with, 130; reversal, apparent, of colors in, 120; slit, mechanism of, 121, 125.
Micrum, 100.
Mikron, 100.
Minerals, colored, absorption spectra of, 134.
Minute objects, arrangement of, 180.

Mirror, 10, 11; for Abbe illuminator, 46; of camera lucida, arrangement for drawing, 112; concave, use of, 36; dark ground illumination, 47; light with, central and oblique, 39; lighting with, 36; plane, use of, 36.
Mixture, clearing, 176.
Moist chamber, 151.
Molecular movement, 88.
Monazite sand, spectrum of, 134.
Mono-refringent, 137.
Mounting, cells, preparation of, 148; media and preparation of, 175 79; objects for polariscope, 137; permanent, 147; temporary 146.
Mounting objects, dry in air, order of procedure, 147; in glycerin, order of procedure, 149; in glycerin jelly, order of procedure, 150; in media miscible with water, 149; minute objects, 180; permanent, 147; in resinous media, 152; in resinous media, by drying or desiccation, order of procedure, 152; in resinous media, by successive displacements, order of procedure, 153; temporary, 146.
Movement, Brownian, or molecular, 88.
Muscae volitantes, 90
Muscular fibers, isolation of, 155.

N

Natural balsam, 176.
Needle-holder, 146.
Negative, aberration, 52; development of, 196; labeling and care of, 197; marking, 197; oculars, 22; rack for drying, 203.
Net-micrometer, 113.
Neutral balsam, 176.
Nicol prism, 136.
Nitrate of uranium, spectrum of, 135.
Nitric acid, dissociator, 179.
Nomenclature of objectives, 11
Non-achromatic condenser, 44; objectives, 12.
Non-adjustable objectives, 13; thickness of cover glass for, table, 15.
Normal salt solution, 179.
Nose-piece, 9, 27; marking objectives on, 65.
Numerical aperture of condenser, 43; objectives, 17, 212; table of, 20; resolution and, 21.

O

Object, determination of form, 83; having plane or irregular outlines, relative position in a microscopical preparation, 82; and image, size of, 5, 11, 96; marking parts of, 63, 64; micrometer, 94 mounting, 146 setting under microscope, 27; shading, 56; suitable for photo micrography, 191; transparent with curved outlines, relative position in microscopic preparations, 83.
Objective, 7, 10, achromatic, 10, 12, 195 adjustable, 12, 13, 52; adjustable, experiments, 51; adjustable, micrometry with, 106; adjustable, photo micrography with, 201, adjustment for, 51; aerial image of, 10; anastigmat, 196; aperture of, 21, 22, 212, aplanatic, 12; apochromatic, 12, 189, back combination of, 12; cleaning back lens of, 58; collar, graduated for adjustment, 53; cloudiness or dust, how to determine, 82, designation of, 11; dry, 11, 17-19; equivalent focus of, 11, 26, 214; field of, 28, 29; focusing for micro-spectroscope, 130; front combination of 10, 11; function of, 29, 30; glass for, 11-13, 63; high, focusing with, 37, homogeneous immersion, 17 19; homogeneous immersion, cleaning, 56, homogeneous immersion, experiments, 51, 55; illuminating, 14; image, power of, 19; immersion, 12, index of refraction of medium in front of, 17, 18; inverted, real image of, 29; for laboratory microscope, 61; lettering, 11; light utilized with, 18, low, attached directly to camera, 198; low, focusing with, 36; magnification of, 214; marking, by Krauss' method, 65; to use with micro polariscope, 136; microscopic, 10; to use with micro-spectroscope, 130; for micro spectroscope, focusing, 130; nomenclature of, 11; non achromatic, 12; non-adjustable, 13; non adjustable, thickness of cover glass for, table, 15; numbering, 11; numerical aperture, 21, 22, 212; oil immersion, 12; panto-chromatic, 13; parachromatic, 13; perigraphic, 196, for photo-micrography, 189; projection, 14, 195; putting in position and removing, 26; semi-apochromatic, 13, table of field, 28; terminology of, 11, unadjustable, 13; variable, 13 visual and actinic foci of, in photo-micrography, 189; water immersion, 17-19, 55; experiments, 51, working distance of, 11, 33, 38.
Oblique light, 35, 39; with Abbe illuminator, 46; experiments, 39, 46, with a mirror, 39.
Ocular, 22; achromatic, 23, Airy's, 23 aplanatic, 12, 23; binocular, 23, cloudiness, how to determine and re-

move, 82 ; Campani's, and cob-web micrometer, 23 ; compensating, 24, 26 ; compound, 23 ; continental, 23 ; deep, 23 ; designation by magnification or combined magnification and equivalent focus, 26 ; diaphragm, 28 ; dust, how to determine, 82 ; equivalent focus of, 26 ; erecting, 23 ; eyepoint of, demonstration, 32 ; fieldlens, 31 ; filar micrometer, 23, 26, 104 ; focus, equivalent of, 26 ; function of, 30, 31 ; goniometer, 23 ; high, 23 ; holosteric, 23 ; Huygenian, 22-24, 31 ; iris diaphragm, 181 ; index, 23 ; Jackson micrometer, 23 ; Kellner's, 23 ; lettering of, 26 ; low, 23 ; micrometer, 23, 25, 102, 105 ; micrometer, micrometry with, 105 ; micrometer ratio, 106 ; table of magnification of, 99 ; micrometer, valuation of, 102 ; micrometer, varying valuation, 105 ; micrometer, ways of using, 105 ; micrometric, 23 ; microscopic, 22, 23 ; negative, 22, 23 ; numbering, 26 ; orthoscopic, 23, 28 ; parfocal, 16, 24, 37 ; periscopic, 24, 28 ; for photo-micrography, 25, 189 ; photo-micrography with and without o., 198, 199 ; pointer, 63 ; positive, 22 ; projection, 25, 189 ; projection, designation of, 189 ; projection, use of in photo micrography, 199, 202 ; putting in position and removing, 27 ; Ramsden's, 24 ; screw micrometer, 26, 103 ; searching, 24, 25 ; shallow, 24 ; solid, 24 ; spectral, 24, 106 ; spectroscopic, 24, 120 ; stauroscopic, 24 ; stereoscopic, 24 ; table of field of, 28 ; working, 25.

Oil, and air, appearances and distinguishing optically, 85 ; removal, 58 ; removal from sections, 161.

Oil-globules, with central illumination, 84 ; with oblique illumination, 85.

Oil-immersion objectives, 12.

Optic axis, 2, 5 ; of condenser, or illuminator, 35 ; of microscope, 6.

Optical center, 2, 3 ; combination, 104, 105 ; focus, 13 ; parts of compound microscope, 10, 61 ; parts of microscope, care of, and testing, 57, 60 ; section, 87

Order of procedure in mounting objects, dry or in air, 147 ; in glycerin, 149 ; in glycerin jelly, 150 ; in resinous media by desiccation, 152 ; in resinous media by successive displacements, 153

Ordinary ray with polarizer, 136.
Orthochromatic plates, 191.
Orthoscopic ocular, field with, 28.
Outline, distinctness of, 85.

Over-correction, 5.
Oxy-hemoglobin, spectrum of, 122, 132.

P

Pantachromatic objective, 13.
Paper, aristotype, 208 ; bibulous, filter, lens, or Japanese, 57, 160 ; bromide, 208 ; for cleaning oculars and objectives, 57, 160 ; funnel, 175 ; Usago, 160.
Parachromatic objective, 13.
Paraffin, 179 ; chloroform, 176 ; chloroform, infiltrating with, 165 ; hard, 166 ; imbedding in, 165 ; method, 164 ; removal from lenses, 58 ; removing from sections, 166.
Parfocal oculars, 16, 24, 37.
Parts, optical and mechanical of microscope, 10, 61 ; testing, 60.
Pedesis, 88 ; compared with currents, 88 ; with polarizing microscope, 89 ; proof of reality of, 89.
Penetrating power, 21, 22.
Penetration of objective, 21.
Perigraphic objective, 196.
Periscopic ocular, field with, 28.
Permanent mounting, 147 ; preparations of isolated cells, 155.
Permanganate of potash, absorption spectrum of, 122, 131.
Picric-alcohol, 179.
Pin-hole diaphragm, 45.
Pipette, 161.
Photo engraving, 216, 219 ; drawing for, 216-219 ; lettering for, 218.
Photographic negatives, marking, 197 ; objectives, 196 ; for photo micrography, 196.
Photography, basis for figures, 206 ; compared with photo-micrography, 183 ; indebtedness to microscopy, 183, 184 ; lighting large objects for, 194 ; objectives for, 196 ; of objects in alcohol or water, 205 ; with a vertical camera, 205, 207.
Photogravures from photo-micrographs, 208.
Photo micrograph, 183 ; developers for, 197 ; determination of magnification for, 202 ; at 5-20 diameters, 193 ; 20-50 diameters, 198 ; 100-150 diameters, 201 ; 50-2000 diameters, 203 ; objects suitable for, 191 ; prints of, 208 ; plates for, 194 ; reproductions of, 208 ; with and without an ocular, 198
Photo-micrographic camera, 186, 188, 190, 192, 193, 202.
Photo-micrography, 183-204 ; cover-glass correction, 201 ; apparatus for, 185 ; compared with ordinary photogra-

phy, 184, condenser for, 42, 191, distinguished from micro-photography, 183; experiments, 193; exposure for, 193, 194, 199, 200, 203; focusing for, 195; focusing screen for, 195; lighting, 192, 193, 198, 203; objectives and oculars for, 14, 189, 195; vertical camera with, 205; visual and actinic foci in, 189; with long and short bellows, 187; with and without ocular, 189-199; record table for, 204.
Physiological histology, 173.
Plane mirror, use of, 36.
Plates, exposure of, 193, 194, 199, 200, 203; gelatino-bromide, 184; isochromatic, or orthochromatic, 189, 191, 197.
Pleochroism, 138
Pleurosigma angulatum, 39
Point, axial, 16; burning, 6.
Polariscope, 127, 136.
Polarized light, extraordinary and ordinary ray of, 136.
Polarizer and analyzer, putting in position, 136.
Polarizing microscope, pedesis with, 89.
Position of objects or parts of same object, 82.
Positive oculars, 11, 22.
Power of microscope, 92; illuminating, penetrating, resolving, 21; of objective, 19, 215; of ocular, 26, 215.
Preparation of Canada balsam, Farrant's solution, glycerin, glycerin jelly, etc. 175-179.
Preparation of clearing mixture, liquid gelatin and shellac cement, 175-179.
Preparations, cataloging, 171, 172; cabinet for, 173 174; labeling, 171; permanent, of isolated cells, 155; storing, 171.
Preparation of diagrams, 215; of ground glass, 29; vials, 159
Price of American and foreign microscope, 61, 63.
Principal focus, 3, 5; focal distance, 3, 30; optic axis, 2, 5.
Prism of Abbe camera lucida, 111; Amici, 126; comparison, 127, 128; dispersing, 126; Nicol, 136; and slit of micro-spectroscope, mutual arrangement, 125; of Wollaston's camera lucida, 111.
Prints and mechanical printing of photo micrographs, 208.
Projection objective, 14, 195; ocular, 25, 26, 189; designation of, 189; in photo micrography, 199, 202.
Putting on cover-glass, 146; an object under microscope, 27; an objective and ocular in position, 26, 27.
Pyroxylin, 176.

Q – R

Quadrant for camera lucida, 113.
Quinine, Herapath's method of determining minute quantities of, 181.
Rack for drying negatives, 203.
Ratio, ocular micrometer, 106.
Reagents for fixing, 175 irrigation with, 159; for mounting, 175.
Real image, 5, 10, 22, 29 31, magnification, 42.
Record table for photo-micrography, 204.
Reflected light, 31.
Refraction, 50; images, 49, 55; index of, 52; of medium in front of objective, 20.
Refractive, doubly, 137; highly, 87; singly, 137.
Relative position of objects, 82.
Resinous media, mounting objects in, order of procedure, by drying or desiccation, 152; by a series of displacements, 153.
Resolution and numerical aperture, 21.
Resolving power, 20.
Retinal image, 6, 10.
Revolver, 27.
Revolving nose-piece, marking objectives on, 65.
Rice, examination of, 90.

S

Sagittal sections, 170.
Salicylic acid, crystallization, 48.
Salt solution, normal, 179.
Scale of drawing, 118; of wave lengths, 128.
Scales of butterflies and moths, examination of, 90.
Screen, color, 191; focusing s. for photomicrography, 194, 195; of ground glass, 29; for microscope, 56.
Screw, society, 62; micrometer, 26, 104.
Sealing cover glass, 148.
Searching ocular, 25.
Secondary axis, 3.
Section, optical, 87.
Sections, arrangement of tissue for, 168, clearing, 167; cutting, 160, 166, fastening to slide, 160, 166, frontal, 170; removing benzin, oil and paraffin from, 166, 161; sagittal, 170, serial, 168 171; staining, 167, 167 transferring, 160.
Sediment in water determination of character, 180
Selenite plate for polariscope, 138.
Semi-apochromatic objective, 13
Serial sections, 168; arranging and labeling, 169, 171 determining thickness

of, 170; stage for, 65; thickness of cover-glass for, 170.
Shellac cement, preparation of, 179; removal from lenses, 58.
Sight, injury or improvement in microscopic work, 59.
Significance of aperture, 20.
Silk, examination of, 90.
Simple microscope, see under microscope, 1.
Slides, 140; cleaning, 140.
Slips, 140.
Slit mechanism of micro-spectroscope, 121.
Society screw, 62.
Sodium, lines and spectrum, 122, 123.
Solar spectrum or s. of sunlight, 122.
Soluble cotton, 176.
Solution, alum, 175; Farrant's, 177.
Spectral, colors, 123; ocular, 120, 126.
Spectroscope, direct vision, 120.
Spectroscopic ocular, 24, 120.
Spectrum, 122; absorption, 123; amount of material necessary and its proper manipulation, 130; analysis, 135; Angström and Stokes' law of, 113; banded, not given by all colored objects, 133; of blood, 131; of carbon monoxide hemaglobin, 133; of carmine solution, 133; of colored minerals, 134; of colorless bodies, 134; comparison, 128; complementary, 124; continuous, 123; double, 128; incandescence, 123; line, 123; met-hemaglobin, 133; monazite sand, 134; nitrate of uranium, 135; oxy-hemaglobin, 132; permanganate of potash, 131; single-banded of hemaglobin, 132; sodium, 122, 123; solar, 122, 123; two-banded of oxy-hemaglobin, 132.
Spherical aberration, 4; test for, 210.
Stage, 61; mechanical, 61, 65, 66; micrometer, 94; for serial sections, 65.
Stain, alcoholic, 161; aqueous, 161.
Staining cells, 155; sections, 161, 167.
Stand, of microscope, 61; for laboratory microscope, 61.
Standard distance (250 mm.) at which the virtual image is measured, 97.
Starch, examination of, 60.
Stokes and Angström's law of absorption spectra, 123.
Storing preparations, 171.
Substage, 67.
Substances for crystallography, 181.
Sulphonal with polarizer, 139.
Sulphuric ether, 177.
Swaying of image, 46.
System, back, front, intermediate, of lenses, 10.

T

Table, for immersion fluid, 213; of magnification and valuation of ocular micrometer, 99; of tube-length and thickness of cover-glasses, 15; natural sines, third page of cover; of weights and measures, second page of cover; of numerical aperture, 20; record, for photo-micrography, 204; size of fields, 28; of valuations of ocular micrometer, 99.
Temporary mounting, 146.
Terminology of objectives, 11.
Test of chromatic and spherical aberration, 210.
Tester, cover-glass, 144, 145; for homogeneous liquids, 55.
Testing a camera, 185; ink, 216; a microscope and its parts, 60.
Test-plate, Abbe's, method of using, 210.
Textile fibers, examination of, 90.
Thickness of cover-glass for non-adjustable objectives, table, 15; of serial sections, 170.
Tissues, arranging for sections, 168; fixing or hardening, 157, 164.
Tolles-Mayall mechanical stage, 65.
Transections, 169.
Transfering drawings, 217; sections, 160.
Transmitted light, 35.
Transparent objects having curved outlines, relative position in microscopic preparations, 83.
Triplet, achromatic, 7.
Tripod, 7; as focusing glass, 198.
Tube of microscope, Frontispiece.
Tube-length, 14-16, 54; for cover-glass adjustment, 53, 54; importance of, 53, 54; microscopical, 14, 15; of various opticians, table, 15; and optical combinations, 104.
Turn-table, 148.

U—V—W—X

Unadjustable objectives, 13.
Under-correction, 5.
Unit of measures, in micrometry, 100; of wave length, 129.
Uranium nitrate spectrum of, 135.
Usago paper, 160.
Valuation of ocular micrometer, 102, 103; table, 99.
Variable objective, 13.
Varying magnification of compound microscope, 97.
Varying ocular micrometer valuation, 105.
Velocity under microscope, 88.

Vertical camera, 188, 190, 192, 193, 202, 207
Vials for preparation 159
Virtual image, 5, 6, 10, 31; standard distance at which measured, 97.
Vision, double or binocular, 92, 93.
Ward's eye shade, 59.
Waste bowl, 162.
Water immersion objective, 17, 19, 55; light utilized, 18; numerical aperture, 20
Water, for immersion objectives, 55; removal, 58; solid sediment in, 180.

Wave length, designation of, 1, 9; scale of, 128
Weights and measures, see 2d p of cover
Wollaston's camera lucida, 65, 111.
Work-room for photo-micrographs, 187
Work-table, position, etc., 59.
Working distance of microscope or objective, 11, 33; determination of, 38; oculars, 25.
Writing diamond, 173.
Xylene, 159, 176; balsam, 175, 176.
Xylol, German form of xylene, 159, 176.

www.ingramcontent.com/pod-product-compliance
Lightning Source LLC
Chambersburg PA
CBHW021354230426
43666CB00006B/519